T0339616

Low-carbon Energy Security from a European Perspective

Low-carbon Energy Security from a European Perspective

Edited by

Patrizia Lombardi
Interuniversity Department of Regional and
Urban Studies and Planning (DIST),
Politecnico di Torino, Turin, Italy

Max Gruenig
Ecologic Institute, Berlin, Germany;
Ecologic Institute US,
Washington, DC, United States

AMSTERDAM • BOSTON • HEIDELBERG • LONDON
NEW YORK • OXFORD • PARIS • SAN DIEGO
SAN FRANCISCO • SINGAPORE • SYDNEY • TOKYO
Academic Press is an imprint of Elsevier

ELSEVIER

Academic Press is an imprint of Elsevier
125 London Wall, London EC2Y 5AS, UK
525 B Street, Suite 1800, San Diego, CA 92101-4495, USA
50 Hampshire Street, 5th Floor, Cambridge, MA 02139, USA
The Boulevard, Langford Lane, Kidlington, Oxford OX5 1GB, UK

Copyright © 2016 Elsevier Ltd. All rights reserved.

No part of this publication may be reproduced or transmitted in any form or by any means,
electronic or mechanical, including photocopying, recording, or any information storage and
retrieval system, without permission in writing from the publisher. Details on how to seek
permission, further information about the Publisher's permissions policies and our arrangements
with organizations such as the Copyright Clearance Center and the Copyright Licensing Agency,
can be found at our website: www.elsevier.com/permissions.

This book and the individual contributions contained in it are protected under copyright by the
Publisher (other than as may be noted herein).

Notices
Knowledge and best practice in this field are constantly changing. As new research and
experience broaden our understanding, changes in research methods, professional practices,
or medical treatment may become necessary.

Practitioners and researchers must always rely on their own experience and knowledge in
evaluating and using any information, methods, compounds, or experiments described herein.
In using such information or methods they should be mindful of their own safety and the safety
of others, including parties for whom they have a professional responsibility.

To the fullest extent of the law, neither the Publisher nor the authors, contributors, or editors,
assume any liability for any injury and/or damage to persons or property as a matter of products
liability, negligence or otherwise, or from any use or operation of any methods, products,
instructions, or ideas contained in the material herein.

British Library Cataloguing-in-Publication Data
A catalogue record for this book is available from the British Library

Library of Congress Cataloging-in-Publication Data
A catalog record for this book is available from the Library of Congress

ISBN: 978-0-12-802970-1

For information on all Academic Press publications
visit our website at https://www.elsevier.com/

Working together
to grow libraries in
developing countries

www.elsevier.com • www.bookaid.org

Publisher: Joe Hayton
Acquisition Editor: Lisa Reading
Editorial Project Manager: Peter Jardim
Production Project Manager: Sruthi Satheesh
Designer: Greg Harris

Typeset by TNQ Books and Journals

Disclaimer: The information, documentation and figures available in this book are written
by the MILESECURE-2050 Consortium partners under EC co-financing (project FP7-
SSH-2012-2-320169) and does not necessarily reflect the view of the European Commission.

Contents

List of Contributors

O. Amerighi Agenzia nazionale per le nuove tecnologie, l'energia e lo sviluppo economico sostenibile (ENEA), Rome, Italy

B. Baldissara Agenzia nazionale per le nuove tecnologie, l'energia e lo sviluppo economico sostenibile (ENEA), Rome, Italy

G. Caiati Laboratorio di Scienze della Cittadinanza (LSC), Rome, Italy

C. Cassen CIRED (Centre International de Recherche sur l'environnement et le développement), CNRS (Centre national de la recherche scientifique), Nogent sur Marne Cedex, France

G. Cotella Interuniversity Department of Regional and Urban Studies and Planning (DIST), Politecnico di Torino, Turin, Italy

S. Crivello Interuniversity Department of Regional and Urban Studies and Planning (DIST), Politecnico di Torino, Turin, Italy

B. Felici Agenzia nazionale per le nuove tecnologie, l'energia e lo sviluppo economico sostenibile (ENEA), Rome, Italy

F. Gracceva Studies and Strategies Unit, ENEA (Italian National Agency for New Technologies, Energy and Sustainable Economic Development), Rome, Italy

M. Gruenig Ecologic Institute, Berlin, Germany; Ecologic Institute US, Washington, DC, United States

M. Karatayev University of Nottingham, United Kingdom

P. Lombardi Interuniversity Department of Regional and Urban Studies and Planning (DIST), Politecnico di Torino, Turin, Italy

B. O'Donnell Ecologic Institute US, Washington, DC, United States

A. Prahl Ecologic Institute, Berlin, Germany

G. Quinti Laboratorio di Scienze della Cittadinanza (LSC), Rome, Italy

G. Valkenburg Maastricht University, Maastricht, The Netherlands

K. Weingartner Ecologic Institute US, Washington, DC, United States, formerly Ecologic Institute, Berlin, Germany

P. Zeniewski Directorate-General Joint Research Centre (DG-JRC), Institute for Energy and Transport (IET), Energy Security Unit, Petten, The Netherlands

Foreword

A Resilient Energy Union with a Forward-looking Climate Change Policy is the third (out of 10) key Juncker Commission priorities. This emblematic policy area builds on several realities, especially the European Union's (EU) heavy dependence from fossil fuels and the EU leadership in the fight against climate change.

More than half of the EU energy supply is covered by third-country imports. Most recent figures show that the EU imports 90% of crude oil and 66% of natural gas, largely from Russian and Middle Eastern producers. The security of EU energy supply requires developing corridors of sustainable and reliable sources at competitive prices. For that purpose, the EU has to speak with one voice when dealing with external energy suppliers in order to reduce its energy consumption and to increase its indigenous energy production.

From Kyoto in December 1997 to the Paris 21st Conference of the Parties in Paris in December 2015, the EU has made important efforts to reduce greenhouse gas emissions (GHGs). Legally binding European legislation has been put in place to achieve the 2020 targets of 20% GHG reduction compared to 1990. Additional 2030 targets have been agreed by EU leaders for reducing GHG by 40%, to increase the share of renewables and increase energy efficiency by at least 27%.

To face these challenges of energy security supply and climate change, the EU has to pool its resources, to combine its infrastructure and to unite its negotiating power vis-à-vis third countries. The EU objectives to become number one in renewable energy, in energy efficiency and in the fight against global warming require actions at the EU, national and local levels in both the medium and long-term.

Long-term and local actions are two particular dimensions covered by the three-year EU research project MILESECURE, the acronym for *Multidimensional impact of the low-carbon European strategy and energy security, and socio-economic dimension up to 2050 perspective*. Funded by the seventh EU Framework Programme (FP7), MILESECURE has been coordinated by Patrizia Lombardi from Politecnico di Torino in collaboration with 10 multidisciplinary research teams coming from Austria, France, Germany, Italy, Netherlands, Poland, the United Kingdom and the European Joint Research Centre.

Low-carbon Energy Security from a European Perspective is an example of good dissemination of research findings. It largely takes stock of MILESECURE evidence, facts and figures but it goes beyond it, complementing the research with key policy dimensions like the geopolitical context and the 21st century world energy order (cf, insights from Giancarlo Cotella, Silvia Crivello and Marat Karatayev), the governance of energy security (Govert Valkenburg and Francesco Gracceva paper) and the relations with EU neighbours and 'large neighbourhood' (see the papers from Gabriele Quinti on North Africa and from Andreas Prahl and Katherine Weingartner on Russia).

MILESECURE researchers are coming from different origins, cultural backgrounds and scientific disciplines. When sociologists argue with engineers, when economists discuss with architects, when political scientists debate with environmentalists, it is a research challenge per se as there is not yet a *lingua franca* for multi/pluri/interdisciplinary projects. These perspectives come from several points in this book, including a historical perspective for European energy-related policies, like in the chapter from Patrizia Lombardi and Max Gruenig on the energy paradigm shift, or in the Commission forward-looking jargon of 'socioecological transition'.

The fact that MILESECURE uses both quantitative and qualitative methods is an additional added value. To translate a vision (like urban sustainability) into a mathematical variable (in the models used) is something highly appreciated and allows science to progress. In this regard, the chapters on low-carbon pathways and on systemic risk-management (cf, Christophe Cassen, Francesco Gracceva, Govert Valkenburg and Peter Zeniewski) are particularly worth reading.

Social acceptability and the role of society in shaping the future energy system are well reflected in the paper on 'anticipatory experiences' by Giovanni Caiatai and Gabriele Quinti. In a nutshell, if big is powerful (cf, coal, nuclear and gas plants), small is beautiful (cf, renewable energy sources and technologies). The many examples of distributed generation and local initiatives all around Europe may well anticipate the EU of 2050 where citizens are capable of creating energy solutions.

To ensure security of energy supply and to remain in the limits of a maximum 2°C increase of temperature compared to the preindustrial basis will be difficult. Strong efforts will have to be done. The EU has committed to reduce its emissions to 80–95% below 1990 levels by 2050. More than half of the world's population live in cities and almost three-quarters will by 2050. The bulk of the increase of urban population (95%) is coming from developing and emerging countries.

Postcarbon cities should aim at reducing vulnerabilities but also at answering to economic and social issues. Resilience is the capacity to successfully deal with difficult events and to adapt or overcome adversity. In a rapidly changing world, resilience creates stability, which in turn promotes environmental sustainability, job creation and economic progress. Win-win-win solutions are technically feasible, economically realistic and socially positive. The 2015 Pope Francis encyclical letter *Laudato si'* underlines the importance of 'integral ecology', a clear message for the current and future generations.

A long-term strategy addressing world, EU, national and local governance issues is needed. But one principle well-identified by MILESECURE should be applied: top-down interventions must be geared towards enabling local initiatives and communities. This is one of the key messages of the MILESECURE manifesto for human-based governance of secure and low-carbon energy transitions.

If Europe, a continent so proud of its roots and shared values, can innovate and co-create sustainably, it will become the beating heart of a new and positive union between individual empowerment and the collaborative economy. Green products and efficient services will be the direct consequences of changing lifestyles providing co-benefits for the economy, the society and the environment.

Domenico Rossetti di Valdalbero
Principal Administrator at the European Commission,
DG for Research and Innovation
Project Officer of MILESECURE

Chapter 1

Challenging the Energy Security Paradigm

M. Gruenig[1,2], P. Lombardi[3], B. O'Donnell[2]
[1]Ecologic Institute, Berlin, Germany; [2]Ecologic Institute US, Washington, DC, United States; [3]Interuniversity Department of Regional and Urban Studies and Planning (DIST), Politecnico di Torino, Turin, Italy

1.1 INTRODUCTION: THE PREVAILING ENERGY SECURITY PARADIGM

The confluence of energy, environment and security policies in the European Union (EU) not only occasion an opportunity to implement comprehensive, ambitious reform agendas for the 21st century, this same nexus also served as the historical foundation of the EU itself. When Robert Schuman, then serving as the French foreign minister, proposed the creation of the European Coal and Steel Community (ECSC) in 1950, securing peace was his primary motivation. Indeed, Schuman envisioned a level of economic cooperation and interdependency that would render a resurrection of hostilities between France and Germany, in the words of the Schuman Declaration, 'not merely unthinkable, but materially impossible' (Schuman, 1982).

1.1.1 Markets Not Munitions

The ECSC was created by the 1951 Treaty of Paris, establishing the first supranational body in postwar Europe with legal-binding authority over domestic policy, specifically in developing a common market for steel and coal. Its focus on natural resources evidenced the importance of environmental regulatory schemes and simultaneously foreshadowed the role the environment would play in coming decades. Moreover, the ECSC's unambiguous coupling of energy consumption with economic development privileged its efforts to substantiate the reconstruction and revitalization efforts, including strategic supply-side subsidies for private enterprises, of the six founding member states (ECSC Treaty, 1951).

This novel 'markets not munitions' approach may have served to prevent armed conflict, but the linking of energy, environment and security, then binding it with economic prosperity, has arguably put in place of war an interminable peace

Low-carbon Energy Security from a European Perspective. http://dx.doi.org/10.1016/B978-0-12-802970-1.00001-2
Copyright © 2016 Elsevier Ltd. All rights reserved.

negotiation predicated, by definition, on historic policy regimes. This inherent backward focus would raise questions as to the ECSC's ability to adapt to social, economic and geopolitical changes. Additionally, mechanisms to evaluate the efficacy of the ECSC's efforts were not well-defined, other than, presumably, the absence of war, making unintentional consequences difficult to assess and even more difficult to mitigate as the Ruhrgebiet, or Ruhr valley, discovered.

The vast supply of coal and steel made the Ruhrgebiet the primary source for Germany's rapid industrial development in the second half of the 19th century and continued its prolific output into the 20th century. It was also a point of constant territorial dispute, particularly between France and Germany, since at least the Franco-Prussian War of 1870. Following occupations and reclamations and two world wars, it was evident to all parties that a longstanding peace would not hold unless the 'Ruhr question' was solved. Following the Potsdam Agreement of August 1945, Britain, supporting a policy of international governance of the Ruhr, and the United States, backing nationalization, convened a bilateral conference in August 1947, in hopes of negotiating a settlement. But it was Schuman's plan with its quasi-neoliberal market approach and supranational system of interdependency that was ratified with the Treaty of Paris (Gillingham, 1987).

As it had done for nearly a century, the Ruhrgebiet continued to fuel the industrial reconstruction and economic recovery of Western Europe during the 1950s. As Michael Hatch points out in his book *Politics and Nuclear Energy: Energy Policy in Western Europe*, coal production in the Ruhrgebiet increased 20% from 1950 to 1957, and employment in the sector rose over 12% (536,800 to 604,000), not including the millions of indirect jobs which benefited from coal's success. In 1968, Hatch notes, as oil's domination of the international energy market became entrenched, the total number employed in the Ruhrgebiet's coal sector plummeted to 272,000. Over the same period, hard coal's share of total energy consumption dropped from 70% to 34%, as the price of oil decreased nearly 60% (Hatch, 1986). Germany spent more than a decade attempting to mitigate the drastic decline of the Ruhrgebiet's industrial strength but was unable to find a solution before Norway's Ekofisk discovery in 1969, which further substantiated the feasibility of competitive oil sourcing in Europe.

The ensuing volatility of energy markets throughout the 1970s and 1980s rendered a long-term revival of the Ruhrgebiet's coal and steel dominance unlikely. The region undertook a far-reaching Strukturwandel, or structural transition, in an attempt to diversify its economic and employment opportunities. Although these efforts have been modestly effective, the Ruhrgebiet, the third-largest urban agglomeration in the EU, continues to be burdened by higher-than-average levels of unemployment (10.4% to national 6.7% in 2014) and is facing extreme demographic change in the coming decades (Bundesamt für Arbeit, Regionalverband Ruhr 2015).

Certainly, international developments were the main driver in the decline of the industrial economy in the Ruhrgebiet, but the correlation, if not the causation, that can be linked to the ECSC's founding should not be disregarded. Did supranationalism impede the Ruhrgebiet's ability to adjust to changing market conditions? Did the market-based approach actually make the coal and steel industries uncompetitive in an international arena? And did the attempts to secure energy in the short-term, relying predominately on old supply regimes, in the end make energy consumers in the six Member States less energy secure?

1.1.2 Expanding the Map

As the ECSC became the model, if not the blueprint, for further European integration, eventually being incorporated under the umbrella European Community (EC) with the European Economic Community (EEC) and the European Atomic Energy Community (EURATOM), the scope of integrative approaches to energy, environment and security policies broadened significantly. In 1991, with the opening of Eastern Europe and the USSR presenting new opportunities for expanding and enhancing energy supply cooperation, the European Energy Charter was developed. The process led to the Energy Charter Treaty of 1994, and, ultimately, to the International Energy Treaty of 2015.

The fundamental principle of the Charter was energy security. As with the ECSC, the Charter would achieve energy security for its members through international, if not supranational, cooperation and liberalization of energy markets. Annexes and amendments were additionally created to promote environmental stewardship, primarily through enhancements in energy efficiency (European Energy Charter, 2015).

Much like Schuman's desire to prevent war between France and Germany through energy policy, the Charter, at least tangentially, sought to bridge the East–West divide that had persisted during the years of the Cold War (Gundel, 2004). Unlike the ECSC, however, the Charter has gone far beyond the borders of Europe, welcoming signatories such as Bangladesh, Mongolia, Niger, and the United States.

The 2015 International Charter pursues a more global agenda, outlining the 'trilemma' of global energy security, economic development and environmental protection (International Energy Charter, 2015). The continued synthesizing of these policy objectives, however, simultaneously reifies the false dichotomy, or trichotomy, that trade-offs are required to achieve the 'ne plus ultra'. But beyond these strategic policy debates is a further impediment to implementing such an ambitious treaty: geopolitical realities.

The Russian Federation signed both the 1991 European Energy Charter and the 1994 Energy Charter Treaty. Its abundance of gas and oil reserves made it an ideal partner for energy importers, particularly those in Eastern Europe. However, having never ratified the treaty and first declaring their intention not to do so in August 2009, Russia was not legally bound to its provisions. The Ukrainian gas crisis, which quickly became a European gas crisis in the winter of 2008–09, was a stark reminder that the fluidity of geopolitical tensions is capable of undermining the Charter's best intentions. (The Russia–Ukraine gas crisis will be discussed in greater detail in Chapter 3.)

1.1.3 Past as Prologue?

Both the ECSC and the Charter have demonstrated an incredible appetite for multilateral cooperation within the EU and internationally for interdependent energy policies. Both have sought to broaden participation through liberalisation of energy markets and peripheral industries and services. And both have been eager to capitalise on moments of geopolitical uncertainty to mollify historical tensions. However, neither the ECSC nor the Charter has fully succeeded in adapting to changing social, economic and political circumstances necessary to provide its members neither with a secure energy future nor with a definitive path toward this end. It is our contention that this is not a consequence of the quality of the agreements but rather of the paradigm within which they were conceived.

The experiences of the Ruhrgebiet and the Russia–Ukraine gas provide concrete evidence of the need for a paradigm shift in the way Europe approaches questions about providing a secure energy future. Can economic development be decoupled from energy consumption? Does geographic expansion of the energy supply chain enhance security? How can energy security be defined and measured, both on the macro and micro levels? And, of particular emphasis for this book, can the success of the EU's climate policies be a guide for, if not wholly incorporate, a comprehensive energy strategy?

The EU is recognized as the world leader in actionable climate policies, a cornerstone of which is promoting the evolution of a low-carbon society. By focussing on this evolution, rather than on historical policy regimes, the EU will be able to provide energy security and continue to serve as a role model for forward-thinking democratic policies.

1.2 HARMONIZING ENERGY AND CLIMATE POLICIES

Energy and climate are intrinsically interlocked: more than 80% of greenhouse gas (GHG) emissions in the EU are energy related (www.eea.europa.eu/data-and-maps/indicators/en01-energy-and-non-energy). Between 1990 and 2012, energy intensity of the European economy decreased on average by 1.7% and the carbon intensity decreased by 9.1% from 2000 to 2012. However, absolute energy consumption remained almost unchanged between 1990 and 2013 in the EU 28 (Eurostat tsdcc320). Moreover, dependency on energy imports rose from 44.3% in 1990 to 53.2% in 2012 (Eurostat tsdcc310).

Thus, in recent years, Europe's energy system has become incrementally lower-carbon and the economy as a whole more energy efficient, while energy dependency increased. Security of supply, sustainability and competitiveness are the three complementary pillars of the European energy policy (COM (2006) 105 final) and have been translated into the main goals of the more recent EU energy strategy (COM (2010) 639 final). These components have also been reconfirmed in the EU's 2020 to 2030 transition framework for climate and energy policies (COM (2014) 15 final).

Policymakers in Europe are faced with the need to achieve energy security and promote a transition towards decarbonised energy sources. However, while the EU has been successful in institutionalising a climate policy, it has not yet been able to formulate a successful energy security policy, in spite of the fact that energy security has been growing in importance in the political agenda. Only a few scenarios, among those produced by key modelling exercises, address the potential synergies between climate change and energy security, and these mainly arise from the agenda of international negotiations over the last 15 years on climate emission reduction targets.

The blurred understanding that still characterises the intersections of energy security and low-carbon society requires further exploration. In particular, the radical changes envisaged by the wide range of policies introduced to pave the way towards a low-carbon energy system may pose challenges to the security of the EU energy system. The EU Green Paper on a 2030 framework for EU climate change and energy policies states that 'the 2030 framework must identify how best to maximise synergies and deal with trade-offs between the objectives of competitiveness, security of energy supply and sustainability' (EC, 2013). The risk is that, if not properly designed, policies aimed at the reduction of GHG emissions may affect the resilience of energy systems and their capacity to tolerate disturbance and to deliver stable and affordable energy services to consumers. Furthermore, by supporting technological and market solutions designed to achieve different policy objectives, climate change policies can have an impact on energy security and generate extra costs. These challenges add to the energy security concerns that climate change already poses to the EU energy system. For example, climate change may have indirect impacts on energy generation and consumption (ie, reduced hydropower production), high electricity and gas demand due to extreme weather conditions or possible accidents related to extreme weather events.

1.2.1 Defining Energy Security

To better understand the current situation, the authors of this book have adopted a definition of a secure energy system as one evolving over time with sufficient capacity to absorb uncertain, adverse events, so that it is able to continue satisfying the energy needs of its intended users, with acceptable changes in volume and price.

Potential threats to energy security are defined from three perspectives: temporality, provenance and society. First, transient disruptions or shocks based on their temporality, such as extreme weather conditions, accidents, terrorist attacks or labour interruptions can be differentiated from more enduring pressures or stresses which compromise the long-term ability to develop adequate physical and regulatory conditions to deliver energy supplies to end-users.

Second, the provenance of threats was defined to allow a distinction between internal and external threats that directly inform the types of strategies that can be put in place for varying situations. The third perspective is the role of society, which is crucial for a secure energy system as part of a transition towards a low-carbon economy. The whole process has to be understood as 'societal', as an organic process that is both the result of intentional actions and the product of the interactions of multiple actors and intended and unintended consequences thereof.

Building on the above, a preliminary review of the current situation of EU and Member States' energy security and low-carbon transition strategies allowed for the identification of three main designations: dependency, consumption and integration.

Dependency describes the EU energy landscape both in terms of trends (market, societal, economic and geopolitical features) and of strategies (actions taken to achieve expected results). It appears that on one hand, the EU is and will continue to be dependent on imports for the majority of its energy needs, with an increase dependency rate of 9% from 1999 to 2009. At the same time, the possibility that a degree of energy independence can be achieved relies both on much higher investments in renewables and on reduced use of less sustainable energy sources (as in the case of nuclear energy, coal and oil). The EU is moving away both from fossil fuels (−5.9% in 10 years since 2000) and from nuclear energy (0.7%), with fuel mixes that vary dramatically in the different Member States. Furthermore, there is an internal issue of dependency within the EU because Member States act differently in the energy market. There is no single 'EU buyer' as countries have different national and internal resources (for example, oil available to countries facing the North Sea), different national and regional policies on the balance between traditional and renewable energy sources in spite of EU addresses and directives, and face market and societal inertia that often hamper efforts towards change.

Consumption exacerbates the dependency of the EU (and of some Member States more than others) in providing energy supplies that may require increased expenditure (as in case of market fluctuations) or complex energy mixes (as in the case of those countries that have to rely almost totally on energy imports). According to the latest available data, the EU is the third major global energy consumer, with 13.4% of total world consumption, behind China and the United States. More consequential, different economic sectors rely on different energy sources and each sector depends heavily on one or a few energy sources. This demonstrates multiple energy dependencies, instead of a diverse mix of energy sources, which might be apparent when only considering macro-level statistics. In addition, within current EU consumption patterns, national situations vary considerably. Specifically, consumption rates require a close consideration of lifestyles, societal organisation of energy use and different patterns of use according to local environmental and cultural conditions. At the local level, for instance, it is possible to see the perverse effects of the lack of a consumption policy in terms of industrial competitiveness and the related phenomenon of 'carbon leakage', or relocalisation due to increased production costs in the EU.

The issues of dependency and consumption would benefit from a coherent strategy able to translate consistent policies across and within the EU.

Integration means improved harmonisation of policies and interventions at different geopolitical levels. Scholars have suggested that we are entering in a new energy world order, in which a country's energy surplus (or deficit) strongly contributes to the determination of a national position in the global system. Recent geopolitical events have drawn attention to the need for energy security while continuing the pathway of decarbonisation of the energy system. While the Ukrainian state crisis revolves to a large degree around European energy supply with Russian natural gas, the global policy community is looking at continuous pressure through extended droughts and extreme weather events such as those seen during the torrential rain falls on the Balkans in 2014, reaffirming the need to mitigate climate change.

The diversification of energy sources and imports and the promotion of self-reliance are at the core of EU energy strategies, but the way in which directives are filtered down to Member States and regions make it difficult to achieve these diversification and self-reliance goals. For instance, although the need for a single, coherent energy market has been made explicit, the process in achieving this is still evolving and the result is far from being achieved.

This book explores these charged topics through insights from a series of novel, new energy project case studies – for example, 'How would a large-scale renewable energy initiative in the North Seas influence the amount of natural gas from the Russian Federation imported to the European Union?' Furthermore, the book demonstrates the need for difficult political conversations, within Europe and beyond, by posing fundamental yet new questions about the energy security paradigm.

1.3 OUR APPROACH

Those studying low-carbon energy security, or those considering its implementation, face a multifaceted challenge and experience that simple solutions are hard to find. This book provides new scientific knowledge on these issues and the general objective of regional, territorial and social cohesion by developing new European models which support and enable energy security at the European, national and local scales. It examines scenarios using multiple perspectives which extend to 2050. This time frame is used to assess the legitimacy and efficacy of policies in terms of the capacity for societies to transition to energy security and to consider the long-term socioeconomic impact of such options.

This book draws on the results of the EU funded collaborative project entitled 'Multidimensional Impact of the Low-carbon European Strategy on Energy Security, and Socio-economic Dimension up to 2050 Perspective' (MILESECURE-2050). The project involved 11 organizations and was funded in part by the European Commission's Directorate General of Research and

Innovation (2013–15). The main results of the project were to bring a socio-economic perspective to low-carbon energy security and energy geopolitics in Europe.

The project has adopted a number of methodological concepts from the transition management theory, the path dependency theory and the vision of creative destruction developed by Schumpeter (1994). Such theories are relevant to examine transitional societal processes based on technological changes and how these changes impact the transitional processes.

Transition management theory is a concept for developing a paradigm shift within a society by leading a gradual, continuous transition from one equilibrium to another (Foxon et al., 2008). Within transition management theory several approaches for examining societal transitions towards energy security exist, such as sociotechnical transitions research, technological innovation systems and coevolutionary dynamics.

Sociotechnical transitions research combines technical, social and historical analysis to examine past and contemporary societal transitions and uses a framework of three different levels: landscapes, sociotechnical regimes and technological niches (Kemp, 1994; Geels, 2002). The technological innovation systems approach differs from the sociotechnical transitions idea in regard to long-term sociotechnical changes in that it focuses on understanding innovation from a systems perspective as opposed to the interaction between technological and social elements. The approach claims that firms and actors innovate mostly in response to incentives coming from the wider innovation system and thus studies feedback mechanisms and interactive relations used in the development and application of new knowledge by science, technology, learning, production, policy and demand. Finally, coevolutionary approaches seek to explain long-term processes of change, claiming that dynamics are determined by casual influences between mutually evolving systems.

In addition to transition management theory, the concept of creative destruction, as visualized by Schumpeter in economic innovation, argues that processes may need disruptive processes of transformation that accompany radical innovation in order to make efficiency gains (Schumpeter, 1994).

Currently, while these concepts are in a process of development, they do not fully explain nor allow for the induction of a societal energy transition. Indeed, in many ways current research places an unequal focus on a limited number of factors, be it the individual, society as a whole, technology, history, political, economic or other factors. A holistic approach to studying societal transition is instead needed (Lombardi, 2014).

This book takes the approach that multiple interrelated and coevolving perspectives (environmental, geopolitical, lifestyle and cultural, political, technological, economic and combined) must be examined to explain possible modes for societal transition. And both present day and historical factors play a critical role. In this context, it is possible that multiple pathways for transition exist or can be created by a number of various combinations.

The authors offer a fundamentally holistic perspective on low-carbon energy security by focussing on:

1. assumptions behind current 'energy security' needs;
2. a new discourse that should develop in response to identifying true societal priorities; and
3. how crucial it is to consider environmental objectives alongside geopolitical and economic interests.

We challenge the existing 'business as usual' perspectives on the energy security topic.

- *Energy security is reactive, rather than creative.* With regard to the energy system, political decisions and investments in our society are made in response to geopolitical conflicts, economic trends and to some extent international treaties. They are not made by envisioning the kind of energy system we really want and need 50 years from now. A major theme of this book is the energy security paradigm and how to reimagine it. The book will suggest the benefit of envisioning energy security through out-of-the-box scenario development and consideration of cultural differences in human behaviour, with respect to the energy system.

- *Policymakers operate in silos.* Environment agencies find it difficult to think about maximising jobs. Similarly, economic planners have a difficult time considering decarbonisation objectives in their regular guidance. A major thread through this book is that it may be quite difficult to achieve all of the objectives in parallel. A process of prioritising objectives is necessary, and such a societal discourse on the topic of energy priorities needs to take place. This book therefore reframes the debate of energy security to a more fundamental question: How important is energy security? What are the aspects of energy security that our society is most interested in achieving – low cost energy? environmental protection? instant access to energy?

- *The term 'low-carbon' does not appear alongside 'energy security'.* Across government agencies, utilities, research institutions and corporations, the energy security topic is considered separate from the renewables and climate topic. While this may be similar to the silo problem identified earlier in this chapter, this has more to do with inertia than decision making. The reality is that the wealth of literature on the supply flows of energy, or market developments in response to changes in energy type, or the environmental impact of fuel switching, for example, have all been simulated and studied under antiquated assumptions of high-carbon activities. Thus, the inertia of research on economic, technological, political, societal and environmental impacts of energy security has only been on fossil technologies. This book will challenge these entrenched perspectives.

1.4 STRUCTURE OF THE BOOK

The authors have developed the following three overarching themes for this book:

1. energy security in a geopolitical perspective
2. reshaping equilibria: renewable energy mega-projects and distributed generation
3. developing policy strategies towards a low-carbon and secure energy system

An overarching narrative is that optimizing the energy system simultaneously across different objectives may be impossible (ie, lowest cost, least environmental impact, minimal downtime, regional supply). The book takes a step back and tries to look at the origins and implications of the concept of energy security which is widely accepted as a given in current policy debates. The first cluster conceptualises the links between energy security and climate policies. It also discusses problems concerning energy in a global scenario and proposes an evaluation of imbalances of energy resources and demand as well as geopolitical tensions between different areas in the world. New energy corridors, physical infrastructures and an exploration of potential regional and macro-regional conflicts (eg, Ukrainian–Russian gas conflict, access to African oil, Middle East conflicts, etc.) are analysed. The second cluster investigates a series of renewable energy (and related) mega-initiatives and projects and experiences with distributed renewable energy, as to their potential influence on energy security, climate change and social acceptance.

Finally, the third cluster introduces a new methodological framework to assess the interactions between energy security and climate policies and finally proposes possible governance strategies for energy security.

In particular, Chapters 2–4 deal with imbalances of energy resources and demand and discusses energy in an international scenario towards a new energy world order, describing future energy geopolitics. It explores the role played by oil and natural gas and particularly the geopolitical relations caused by the latter with specific reference to current geopolitical energy situation (ie, the Persian Gulf, the Caspian Sea and Africa). Finally, it highlights present challenges and future perspectives for energy production and consumption in the European Union.

The Chapters 5 and 6 illustrate a number of case studies related to energy security and renewable energy in Europe and outside European boundaries, concluding with lessons learnt during the last decade regarding climate change mitigation, costs and energy security, and how to balance that with competitiveness concerns in Europe.

The Chapters 7–9 are about developing policy strategies towards a low-carbon and secure energy system. An alternative framework is proposed to move along the axis of knowledge, evaluating the co-benefits of climate policies on energy security challenges and finally suggesting a new governance of energy security and climate change mitigation.

The final concluding chapter will revisit key themes and suggests further research in the field.

REFERENCES

European Coal and Steel Community - ECSC, Treaty Establishing the European Coal and Steel Community, April 18, 1951. p. 261U.N.T.S. 140, Paris.

European Commission COM (2006) 105 final. Green Paper on a European Strategy for Sustainable, Competitive, and Secure Energy. European Commission, Brussels.

European Commission COM (2010) 639 final. Communication from the Commission to the European Parliament, the Council, the European Economic and Social Committee and the Committee of the Regions. Energy 2020. A Strategy for Competitive, Sustainable and Secure Energy. SEC 1346, Brussels.

European Commission COM (2013) 150 final. Green Paper Long-term Financing of the European Economy (Text with EEA Relevance), SWD 76 final. Brussels, 25.03.13.

European Commission COM (2014) 15 final. Communication from the Commission to the European Parliament, the Council, the European Economic and Social Committee and the Committee of the Regions. A Policy Framework for Climate and Energy in the Period from 2020 to 2030, (Brussels).

European Energy Charter (1991), in the publication on the Energy Charter Treaty and Related Documents, 17 December 1991. At: http://www.energycharter.org/process/energy-charter-treaty-1994/energy-charter-treaty/.

European Energy Charter and International Energy Treaty in 'The International Energy Charter Consolidated Energy Charter Treaty (With Related Documents)', June 12, 2015.

Eurostat Tsdcc310. Available at: ec.europa.eu/eurostat/product?mode=view&code=tsdcc310.

Eurostat Tsdcc320. Available at: ec.europa.eu/eurostat/product?mode=view&code=ten00086.

Foxon, T., Gross, R., Pearson, P., Heptonstall, P., Anderson, D., 2008. Energy Technology Innovation: A Systems Perspective, Report for the Gaunaut Climate Change Review. Imperial College Centre for Energy Policy and Technology, London.

Geels, F.W., 2002. Technological transitions as evolutionary reconfiguration processes: a multilevel perspective and a case-study. Research Policy 31, 1257–1274.

Gillingham, J., 1987. Die französische Ruhrpolitik und die Ursprünge des Schuman-Plans. Eine Neubewertung. Vierteljahrshefte für Zeitgeschichte 35 (1), 1–24.

Gundel, J., 2004. Regionales Wirtschaftsvölkerrecht in der Entwicklung: Das Beispiel des Energiecharta-Vertrages. Archiv des Völkerrechts 42 (2), 157–183.

Hatch, M.T., 1986. Politics and Nuclear Power: Energy Policy in Western Europe. University Press of Kentucky, Lexington, Kentucky.

International Energy Charter, 2015. http://www.energycharter.org.

Kemp, D., 1994. Global Environmental Issues: A Climatological Approach, second ed. Routledge, London, New York, pp. xvi+224.

Lombardi, P., 2014. Local experiences in energy transition. In: Lcs-rnet Transition and Global Challenges towards Low Carbon Societies, In: EAI: Energia, Ambiente e Innovazione, anno 61, pp. 55–59 Available at: http://www.enea.it/it/pubblicazioni/pdf-eai/speciale-transition-and-global-challenges/eai-speciale-i-2015.pdf.

MILESECURE-2050-Multidimensional Impact of the Low-Carbon European Strategy on Energy Security, and Socio-Economic Dimension up to 2050 perspective.

Schuman, R., 1982. The Schuman Declaration. Selection of Texts Concerning Institutional Matters of the Community from 1950 to 1982. European Parliament - Committee on Institutional Affairs, Luxembourg.

Schumpeter, J.A., 1994. Capitalism, Socialism, and Democracy. Routledge, London.

Chapter 2

European Union Energy Policy Evolutionary Patterns

G. Cotella[1], S. Crivello[1], M. Karatayev[2]
[1]*Interuniversity Department of Regional and Urban Studies and Planning (DIST), Politecnico di Torino, Turin, Italy;* [2]*University of Nottingham, United Kingdom*

2.1 INTRODUCTION

Europe's uneven geographies of energy play a prominent role in influencing the European Union (EU) agenda. For at least 30 years they constitute the object of a growing set of strategies and policies aiming at energy security as well as the transition towards a low-carbon society. These challenges are even more crucial if one considers that their main trends have been dramatically changed by the current economic crisis, which has forced to reshape growth scenarios and readapt expected societal changes.

In this light, this chapter focuses on the main trends in EU climate and energy policy. To do so, the following text is organised into three main sections. The first part takes a historical look at the development of the EU energy policy over the last 50 years, particularly focussing on such policy drivers as the economic integration, the first oil crisis, environmental pollution and climate change. The second part looks at the current energy regimes of the EU and its Member States, focussing on the capacity to reduce their dependency on external energy sources, to improve the energy mix favouring renewables and to pursue a sustainable model of development. The third section turns to the main strategies that have been approved at the EU level to favour the transition from present energy system to a more sustainable and secure energy system. The section considers EU renewable and energy efficiency policy, environmental and climate protection policy, market integration and smart-grid networks, and innovation and technology programmes. The chapter ends by summarising the main challenges that still needs to be tackled by EU climate change and energy policy.

2.2 THE DEVELOPMENT OF THE EU ENERGY POLICY

Energy lies at the heart of the EU integration process. The Treaty of Paris, that in 1951 established a European Coal and Steel Community (ECSC) consisting

Low-carbon Energy Security from a European Perspective. http://dx.doi.org/10.1016/B978-0-12-802970-1.00002-4
Copyright © 2016 Elsevier Ltd. All rights reserved.
13

of France, West Germany, Italy and the three Benelux countries, clearly proves this argument, aimed as it was at favouring European diplomatic and economic stability after World War II. The establishment of a European Atomic Energy Community (EURATOM) in 1957, alongside the European Economic Community (EEC) agreed upon through the Treaties of Rome, reinforced this purpose, aiming at developing nuclear energy in order to fill the deficit left by the expected exhaustion of European coal deposits, to reduce dependence on foreign oil producers and to create a common market for nuclear power in Europe.

Despite its forward-looking solidarity principles, however, the European Community's competence to act in the energy field remained quite limited throughout the 1960s and the first half of the 1970s, with the Member States continuing to hamper the possibility for the EU to speak with one voice vis-à-vis the external energy producer and supplier countries, and to design EU-wide energy policies. These were years when the seven greatest oil companies controlled most of worldwide oil production (82% of the total), refining (65%) and distribution (62%). During that period, the countries belonging to the Organisation of the Petroleum Exporting Countries (OPEC) were responsible for most of oil production worldwide (EC, 2013).[1]

2.2.1 The Arab–Israeli War and the Oil Crisis

The situation forcedly changed in the 1970s, with the Arab oil embargo started in 1973 in the aftermath of the Yom Kippur War between Egypt, Syria and Israel. European oil supplies were cut by 10% on average, bringing about fourfold oil price increases and leading to the exponential growth of inflation rates throughout the 1970s. This caused the worst economic recession in Europe since World War II. The origins of the oil crisis can be explained by the simultaneous presence of a multiplicity of geopolitical and economic factors, threatening the stability of the energy system. This situation generated the emergence of so-called 'energy nationalism', ic, the increasing political will of various countries to autonomously manage their own natural resources and to control directly oil production and distribution. As a consequence, the international oil market changed abruptly, with the entry of state actors that determined the crisis of the prevailing oligopolistic system.[2]

In reaction to the EU and US support of Israel, OPEC countries cut their oil supply to Western countries. Whereas the general opinion of experts was that

---

1. The oil produced by OPEC countries was directed mainly to the EU and US markets (68% of total production), followed by Asian countries (12%), developing countries (12%) and other OECD countries (8%).
2. The foundation of OPEC in 1960, the Tehran agreement in 1971 between six oil exporting countries, and the subsequent Tripoli agreement led to price controls and to the imposition of fiscal rules to international oil companies. As a consequence of these events, the demand–supply relationship on the world crude oil market changed substantially.

the wide availability of oil reserves worldwide would have been able to satisfy an exponentially rising demand, the 1973 facts moved the focus away from the geological availability of oil reserves to the geopolitical conditions and prices of oil supply from instable areas such as the Middle East. The political reaction of oil importing countries to such external events concerned the adjustment of the oil market through a series of measures aimed at improving the short-term resilience of the energy system to sudden external shocks as well as its medium to long-term robustness.

The political reaction of the EU, however, was very fragmented. In this situation of high demand, no spare capacity and insecurity as to the Arabs' intentions, self-help behaviour was at its highest (Chakarova, 2010). EU Member States implemented measures that, together with the lack of coordination, created additional distortions on the oil market. In particular, they pursued bilateral deals with oil producing countries, introduced export restrictions and decided to increase their oil reserves (stockpiling), thereby exacerbating the pressure on the market in a period of limited supply.[3]

On the other hand, during the same years various attempts were put in place by supranational organisations aiming at improving international cooperation in the energy field as well as at reaching a common view across Member States on strategic energy issues. In 1974, the International Energy Agency (IEA) was established to represent oil producer and consumer countries within the OECD members. This was meant to create an energy alternative in order to counterbalance the market strategies and geopolitical power of OPEC countries. On the European side, the Council Resolution concerning a new energy policy strategy for the Community was approved in the same year, underlining the added value of close coordination among Member States to tackle energy problems and laying down guidelines concerning energy supply and demand, notably for 'reduction of the growth of energy consumption and improvement of security of supply under the most satisfactory economic conditions possible'. This energy strategy focused on the promotion and development of alternative, preferably indigenous, energy sources, on the diversification of external energy suppliers and on the promotion of a change in energy use by all stakeholders (rational use of energy, energy conservation measures, etc.).

2.2.2 Chernobyl and the Rise of Environmental Concerns

The energy sector has been subject to major accidents and catastrophes that have marked its history. The control of the environmental impact of the various energy systems in terms of emissions, wastes and perturbation of ecosystems,

---

3. France was the most active in signing bilateral contracts with oil producers. This might have been due to their higher dependence on oil (62% of the energy mix) as compared to other European countries (such as Germany, 53%, and the UK, 44%). France turned out to be also the most fervent opponent to common action within the EEC or the OECD.

under normal or accident operating conditions is hence a major issue. The Chernobyl nuclear accident certainly represents a prominent example in this sense as, together with the Three Mile Island accident in the United States (1979), it is one of the most serious accidents in the history of nuclear power generation.

In 1986, the world industry of nuclear power for civilian purposes consisted of 366 functioning power plants, mostly concentrated in five countries (United States, Japan, France, Soviet Union and Western Germany) that together owned 72% of the entire world generation capacity. As discussed above, the nuclear power option had come out as one of the EU energy policy response to the 1973 oil crisis. In the five years preceding the Chernobyl accident, however, the construction of nuclear power plants in Europe had slowed down because of a set of different factors that changed the national energy strategy adopted by the single countries. The Chernobyl accident therefore only accelerated a process of decreasing reliance on nuclear power for civilian purposes. With respect to the financial, economic and political problems considered as relevant at that time, the accident turned the attention of European public opinion towards environmental protection and human health and safety issues.

The policy response of EU Member States to the rising pressure by domestic public opinion and environmentalist groups proved to be quite heterogeneous and mostly concerned the short to medium time period. One year after the Chernobyl accident, Italy banned nuclear power generation within its own frontiers as a national referendum. The German government launched a policy plan aimed at improving the security and safety of existing installations. France, strongly dependent on nuclear power to meet its energy needs, maintained, but in reduced forms, its programme for the construction of new nuclear power plants. In the 1990s, several other EU members, including Spain, the Netherlands, Sweden and Belgium, opted to force the early closure of existing nuclear plants. In general, the EC took a neutral view of nuclear power leaving the decision on whether or not to use it to the individual member states (Bahgat, 2011).

Following the Chernobyl accident, the issue of environmental protection became more prominent in the EU. This, however, did not immediately lead to including energy issues explicitly into European legislation, especially as climate change was not yet high on the agenda. In 1987, the Single European Act stressed the role of the Community 'to preserve, protect and improve the quality of the environment, to contribute towards protecting human health, and to ensure a prudent and rational utilization of natural resources' in the spirit of 'subsidiarity'. In 1992, the Treaty of Maastricht, while consolidating the European economic integration process, did not contain a separate energy chapter, with energy being mentioned under the 'Environment' title. Similarly, the Treaty of Amsterdam (1999) and the Treaty of Nice (2003) did not bring any major advances for a common energy policy either. The directives on the electricity and gas internal markets issued in 1996 and 1998 were motivated by economic objectives as well. All important energy regulation adopted in the following years, such as the Renewables Directives and the introduction of emissions trading in 2005, were based instead on environmental regulation. At the same time a common European view of energy issues remained latent.

2.2.3 The Kyoto Protocol: Towards a Global Approach to Environmental Problems

While energy has been a pillar of European policy since its inception, concerns about climate change recently have radically changed the European energy policy framework. A milestone in the history of the climate change debate occurred in 1988, when the Intergovernmental Panel on Climate Change (IPCC) was established by the United Nations Environment Programme (UNEP) and the World Meteorological Organisation (WMO), to provide the world with a clear scientific view on the current state of knowledge in climate change and its potential environmental and socioeconomic impacts.

The First Assessment Report (FAR) of the IPCC, published in 1990, predicted temperature rises by the end of the 21st century of 3°C, accompanied by a sea level rise of 65 cm, confirming that climate change is an actual threat to human living conditions and stimulating the international community to act (IPCC, 1990). The first international policy response in 1992 was the adoption of the United Nations Framework Convention on Climate Change (UNFCCC) negotiated at the so-called Earth Summit held in Rio de Janeiro. The ultimate objective of the UNFCCC was to 'stabilize greenhouse gas concentrations in the atmosphere at a level that would prevent dangerous human interference with the climate system' (UNFCCC, article 2). Above all, the UNFCCC sets an overall framework for intergovernmental efforts to tackle the challenge posed by climate change, recognising that the climate system is a shared resource whose stability needs a global approach.

In 1997, binding obligations for developed countries to reduce their greenhouse gas (GHG) emissions were set, and climate change and thus energy issues gained prominence on the global agenda. In this way, policymakers explicitly acknowledged that climate change is of such a scale that solutions cannot be found on the nation-state level. Within this context, the EU played a key role: in 2007, European leaders agreed on the reduction of GHG emissions by at least 20% by 2020. The EU wants to reach this ambitious objective using efficiency measures to reduce total energy consumption, an extended emission trading system in order to give incentives to reduce CO_2 emissions and by increasing the share of energy from renewable sources, while at the same time ensuring international competitiveness and security of supply. Climate change objectives and energy security are therefore currently key drivers for future European energy policy. Policymakers are now under increasing pressure to address these twin challenges: to develop cost-effective policies that will both ensure the security of our energy system and reduce GHG emissions (IEA, 2007).

2.3 CURRENT ENERGY REGIME AND MAIN CHALLENGES FOR ENERGY SECURITY

2.3.1 EU Energy Consumption and Dependence

More recent studies identify three essential characteristics of existing EU energy regime. First, the EU's energy system is characterised by a high rate of energy consumption. Gross inland consumption of energy within the EU in 2013 was 1.666 Mtoe, accounted for 16% of global energy consumption (see Table 2.1). Having remained relatively unchanged during the period from 2003 to 2008, gross inland consumption of energy decreased by 5.7% in 2009 in particular due to the economic recession, climatic conditions and energy efficiency improvements. Germany had the highest level of gross inland consumption of energy in 2013, accounting for a 19.5% share of the EU total. France (15.6%) and the United Kingdom (12.1%) were the only other EU Member States to record double-digit shares, with Italy's 9.6% share just below this level. Together these four Member States accounted for 56.7% of the EU's gross inland consumption.

Second, the EU members have the highest rate of import dependence. The EU imports 53% of the energy it consumes including almost 90% of its crude oil, 66% of its natural gas and 42% of its solid fuels such as coal. In 2013 the bill for external energy amounted to about €400 billion. Thus, access to secure and affordable supplies of fossil fuel remains essential to the economic propriety of EU countries. This is valid in particular for less integrated and connected regions such as the Baltic and Eastern Europe. The most pressing energy security of supply issue is the strong dependence from a single external supplier. This is particularly true for gas, but also applies to electricity. Six Member States depend on Russia as single external supplier for their entire gas imports and three of them use natural gas for more than a quarter of their total energy needs. In 2013 energy supplies from Russia accounted for 39% of EU natural gas imports or 27% of EU gas consumption; Russia exported 71% of its gas to Europe with the largest volumes to Germany and Italy. For electricity, three Member States (Estonia, Latvia and Lithuania) are dependent on one external operator for the operation and balancing of their electricity network. The third essential characteristics of EU energy regime is that fossil fuels continued to dominate primary energy consumption in the EU, but their share declined from 82.1% in 1990 to 72.9% in 2013. The share of renewable energy sources more than doubled over the same period, from 4.5% in 1990 to 12.6% in 2013, increasing at an average annual rate of 4.5% per year. The share of nuclear energy in gross inland energy consumption increased slightly from 13.1% in 1990 to 14.4% in 2013. However, within the EU there are significant differences in energy mix, particularly in terms of the role of coal, gas and nuclear power and in some cases renewable energy. For example, natural gas accounted for 36.5% in the United Kingdom. In France, nuclear power accounted for 45% of total primary energy demand, and in Norway hydropower's share was 65%. The

TABLE 2.1 Gross Inland Consumption of Energy, 1990–2013 (EC, 2013)

	1990	1995	2000	2005	2010	2011	2012	2013	Share in EU, 2013 (%)
EU	1667.3	1671.1	1725.8	1824.7	1760.5	1698.1	1686.1	1666.3	100
Germany	366.3	341.6	342.3	341.9	333.1	316.7	318.6	324.3	19.5
France	227.8	241.8	257.5	276.7	267.6	258.1	258.3	259.3	15.6
United Kingdom	210.6	222.3	230.6	234.6	212.2	198.1	203.1	201.1	12.1
Italy	153.5	151.8	174.2	187.5	174.8	172.1	166.3	160.1	9.6

leader in terms of renewables is Denmark, which accounted for 23% of total primary energy.

2.3.2 The Challenges for Energy Security

The framework presented above poses various challenges in relation to EU energy security (Young, 2011), as for instance the need to import energy sources from external countries and the related risks of a partial or total interruption of physical energy flow (physical risk) and/or a major shift in energy prices (economic risk) (Quemada et al., 2011). This is the reason why, as it will be further elaborated in Section 2.4, the differentiation of energy sources and energy imports and the promotion of self-reliance lay at the core of European energy policy.

European imports of energy come from a limited number of key territories, whose relative importance varies from source to source. The geography of European energy dependency shows the pivotal role of Russia as the one main partner providing oil, natural gas and solid fuels. This makes the EU–Russia relation highly strategic, as testified by the institutionalisation of the EU–Russia Energy Dialogue in Paris in 2000 with the remit of enabling 'progress to be made in the definition of an EU-Russia energy partnership and arrangements for it' (Averre, 2010).[4]

Secondly, it has to be considered that OPEC countries, all together, represent the region of origin of 36% on European oil imports, surpassing Russia (EC, 2011). High oil prices and uncertainty in oil markets have been encouraged, in many periods, by growth in oil demand and a reduction of spare capacity. In this context, the EU and OPEC have established in the second half of 2004 a bilateral dialogue to enhance producer–consumer relations. In this concern, the EU surely aims at a more stable international oil market and prices, an attractive investment climate, more transparency, a better market analysis and forecasts as well as technological and international cooperation.

The price of energy in the EU depends on a range of different supply and demand conditions, including the geopolitical situation, import diversification, network costs, environmental protection costs, severe weather conditions, and levels of excise and taxation. In this sense, maintaining stable prices is a complex issue. At the same time, the price and reliability of energy supplies, electricity in particular, are key elements in a country's energy supply strategy. Electricity prices are of particular importance for international competitiveness, as electricity usually represents a significant proportion of total energy costs for industrial and service-providing businesses. In contrast to the price of fossil fuels, which are usually traded on global markets with relatively uniform prices, there is a wider bandwidth of prices within the EU Member States for electricity or natural gas. The price of electricity and natural gas is also, to some degree,

---

4. EU–Russia Energy Dialogue has been both the first EU strategic energy policy dialogue established with an external energy partner as well as the first sectoral dialogue with Russia (EC, 2011).

influenced by the price of primary fuels and, more recently, by the cost of carbon dioxide (CO_2) emission certificates.[5]

Finally, as it will be discussed later in this chapter, a key strategy for both energy security and environmental sustainability is the promotion of renewable energy sources. Today, European energy mix relies on renewable sources for 10–13% (according to different statistical sources) of its total energy consumption. More renewable energy will enable the EU to cut GHG emissions and make it less dependent on imported energy, and boosting the renewables industry will encourage technological innovation and employment in Europe.

2.4 PAST, PRESENT AND FUTURE EUROPEAN UNION CLIMATE AND ENERGY STRATEGIES

2.4.1 The Consolidation of Contemporary European Energy Strategies

The first significant initiative in this sense was, in 2000, the Green Paper entitled *Towards a European Strategy for the Security of Energy Supply* (EC, 2000). According to the Green Paper, the main objective of an energy strategy should be to ensure, for its citizens and for the proper functioning of the economy, the uninterrupted physical availability of energy products on the market at an affordable price for all consumers, whilst respecting environmental concerns and looking towards sustainable development.

In 2006, a second Green Paper called *A European Strategy for Sustainable, Competitive and Secure Energy* (EC, 2007) outlined three main objectives for the EU common energy policy, namely:

- *Sustainability*, to be achieved through actions including the development of competitive renewable energy sources, the diffusion of alternative transport fuels, the curbing of energy demand within Europe, the development of global actions to halt climate change and improve local air quality.
- *Security of supplies*, which means tackling EU's rising dependence on imported energy through integrated approaches. Possible approaches includes the promotion of demand reduction, the support for growing diversification of EU's energy mix, with greater use of indigenous and renewable energy sources, the geographical differentiation of importers and the promotion of investments in energy-efficient technologies.
- *Competitiveness*, implying the promotion of a functional, open and competitive internal energy market in order to allow improvements in the efficiency of

---

5. One should notice that energy prices are not strictly connected only to phenomena taking place outside Europe. A relevant impact is exerted by the market distortion caused by different prices of energy in different countries, this being the reason for the EU aiming at the institution of a single, coherent energy market (see EC, 2012).

energy grids, a decrease in energy prices, an increase in investments in clean energy production and energy efficiency and an overall improvement of the EU economy in the global scenario.

This document marked the beginning of a more integrated EU energy policy – with the introduction of the concept of 'solidarity' across Member States – and somehow represented the first EU 'energy action plan'. In fact, it identified six key areas where action was required to tackle energy challenges: (1) competitiveness and the internal energy market; (2) diversification of the energy mix; (3) solidarity; (4) sustainable development; (5) innovation and technology and (6) external policy.

The progress of the energy sector in these six areas was to be monitored and enhanced through regular Strategic Energy Reviews. In particular, the *Second Strategic Energy Review: An EU Energy Security and Solidarity Action Plan* (EC, 2008) tried to set up comprehensive long-term strategies in the energy sector with the aim of diversifying the energy mix and sources and of developing EU's indigenous resources. It focused on five major themes: (1) promoting infrastructure essential to the EU's energy needs; (2) a greater focus on energy in the EU's international relations; (3) improved gas and oil stocks and crisis response mechanisms; (4) new impetus on energy efficiency and (5) better use of the EU's indigenous reserves (Fig. 2.1).

Nevertheless, it was since 2007 that several legislations laid out quantitative targets for EU energy polices with the main objective of changing the European energy system to a low-carbon economy, ensuring safe, secure, affordable and sustainable energy supplies. Among them, the *Strategic Energy Technology Plan Towards a Low Carbon Future* provides a framework to accelerate the development and deployment of cost-effective low-carbon technologies including

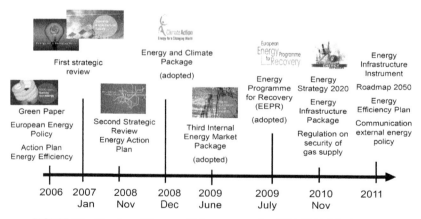

FIGURE 2.1 Timeline of European Union energy policy 2006–2011 (Vinois, 2010).

renewable electricity, second-generation biofuels, smart grids, carbon capture and storage, among others. In addition, in September 2007 the European Commission (EC) adopted the third Internal Energy Market legislative package, aimed at ensuring effective competition and creating the conditions for investment, diversity and security of supply. The objective was to make the energy market fully effective and to create a single EU gas and electricity market keeping prices as low as possible and increasing standards of service and security of supply.

More recently, several strategy papers have been defining further energy developments at the EU level. On March 2010, at the Spring Summit, EU heads of state approved the EC's proposal for the *Europe 2020* economic strategy, the EU's growth strategy for the coming decade. Replacing the much-criticised Lisbon strategy, *A European 2020 Strategy for Smart, Green and Inclusive Growth* continues to promote EU growth based on knowledge and innovation, aiming at achieving employment and social cohesion in a sustainable perspective (Groenenberg et al., 2008; EEC, 2013). The goal of the strategy is to address the weaknesses of the traditional EU growth model in order to create the conditions for a different kind of growth that is smarter (through more effective investments in education, research and innovation), more sustainable (thanks to a decisive move towards a low-carbon economy) and more inclusive (with a strong emphasis on job creation and poverty reduction). To this end, the strategy includes five headline targets to be reached by 2020: employment; education; research and innovation; social inclusion and poverty reduction; and climate and energy. According to the EC, these targets have to be realised through measures based on EU Emissions Trading System (EU ETS), national targets for non-EU ETS Emissions (Effort Sharing Decision), national renewable energy targets and carbon capture and storage (Fisher, 2012).

The European Council, in line with positions of world leaders in the Copenhagen and the Cancun Agreements, has also defined a long-term strategy for decarbonisation. In order to keep climate change below 2°C, the European Council reconfirmed in February 2011 the EU objective of reducing GHG emissions by 80–95% by 2050 compared to 1990, in line with IPCC recommendations.

Analysing in more detail the *Energy 2020: A Strategy for Competitive, Sustainable and Secure Energy* approved in 2011, it appears clear how the document, while putting forward ideas that could form the basis for a new European energy policy, extensively builds on the three core pillars already identified in the 2006 Green Paper (sustainability, security of supplies and competitiveness [Chester, 2010]). In this perspective, the strategy focuses on five priorities:

1. *Achieving an energy-efficient Europe.* Efforts to improve energy efficiency should consider the whole energy chain, from energy generation and transformation to distribution and consumption. Energy efficiency may be supported through technical innovations and investments, and it also concerns the empowerment of domestic and business consumers, their involvement in sustainable choices concerning energy savings, reduction of

wastages and the switching to low-carbon technologies and fuels. Market instruments as emissions trading systems and taxation systems may be of great help in this framework (Rio, 2012).

2. *Promoting a truly pan-European integrated energy market.* The hypothesis is that an integrated, interconnected and competitive market will lead to improvements in efficiency. Currently, Europe is largely fragmented into separate national energy markets characterised by monopoly or oligopoly (Glachant and Lévêque, 2009; Birchfield and Duffield, 2011). In order to promote market integration, the Commission has identified 12 priority corridors and areas, and 248 key energy infrastructure projects concerning energy transport networks. Specific emphasis is posed on the Southern Gas Corridor as natural gas will probably continue to play a key role in the EU's energy mix in the coming years and gas can also gain importance as the backup fuel for variable electricity generation (Finon, 2011; Bilgin, 2009). At the same time, smart metres and power grids are crucial for promoting renewable energies, energy savings and improvements in energy services (Faruqui et al., 2010; Clastres, 2011). Finally, construction of new interconnections at Europe's borders should receive the same attention and policies as intra-EU projects in order to ensure the stability and security of supplies.

3. *Empowering consumers and achieving the highest level of safety and security.* It is crucial to guarantee trust, protect consumers and to help them play an active role (Kolk, 2012). As energy, in particular electricity, constitutes a substantial part of the total production costs of key European industries, it is crucial to provide affordable but cost-reflective and reliable supplies (Bunse et al., 2011). Safety nets are also necessary, for example, for vulnerable consumers. Promoting cooperation, competition and common regulation between Member States can also contribute to diversification of supply sources. The internal market is also hampered when Member States are not fully interlinked, such as in the Baltic States (Roos et al., 2012). Energy policy is also responsible for protecting European citizens from the risks of energy production and transport. For example, a controversial topic is nuclear power (Visschers et al., 2011): currently, mainstream EU position is that Europe must continue to be a world leader in developing systems for safe nuclear power, the transport of radioactive substances, as well as the management of nuclear waste (Dittmar, 2012). International collaboration on nuclear safeguards plays a major role in ensuring nuclear security and establishing a solid and robust nonproliferation regime. In the oil and gas exploitation and conversion sector, the EU legislative framework should guarantee the highest level of safety and an unequivocal liability regime for oil and gas installations.

4. *Extending Europe's leadership in energy technology and innovation.* Decarbonising is strictly connected to technological shifts (Schmidt et al., 2012; Nilsson and Rickne, 2012). The EU ETS is an important demand-side

driver supporting the deployment of innovative low-carbon technologies (Kemp and Pontoglio, 2011). New technologies will reach markets more quickly and more economically if they are developed through collaboration at the EU level. Europe-wide planning and management is therefore paramount for investment stability, business confidence and policy coherence. The Strategic Energy Technology (SET) Plan sets out a medium-term strategy. Main technologies to be developed concern second-generation biofuels, smart grids, smart cities and intelligent networks, carbon capture and storage, electricity storage and electro-mobility, next-generation nuclear, renewable heating and cooling (Fisher, 2012). The resources required in the next two decades for the development of these technologies are very significant, especially when seen in the context of the current economic crisis (Wüstenhagen and Menichetti, 2012). The EU is facing fierce competition in international technology markets; countries such as China, Japan, South Korea and the United States are pursuing an ambitious industrial strategy in solar, wind and nuclear markets (Timilsina et al., 2012).

5. *Strengthening the external dimension of the EU energy market.* As emphasised in the communication, 'The EU Energy Policy: Engaging with Partners Beyond Our Borders', the Commission supports intergovernmental energy agreements between Member States and third countries in order to pursue security of supply, competitiveness and sustainability (Young, 2011). Both relations with producing and transit countries, and relations with large energy-consuming developing countries are of great importance. The EU already has a series of complementary and targeted frameworks ranging from specific energy provisions in bilateral agreements with third countries (free trade agreements, partnership and cooperation agreements, association agreements, etc.) and memoranda of understanding on energy cooperation, through to multilateral treaties such as the Energy Community Treaty and participation in the Energy Charter Treaty. At the same time, more effective coordination at the EU and Member State level is needed. Of course, the external dimension of EU energy policy must be coherent with other external dimensions concerning development, trade, climate and biodiversity, enlargement, common foreign and security policy, etc.

Complementarily, the EC also developed a long-term energy perspective through the *Energy Roadmap 2050*. This document has a longer timeframe giving a direction for EU energy policy after 2020, with the aim of reducing pollutant emissions in 2050 by 80–95% compared to 1990 levels. *Energy Roadmap 2050* explores the way of achieving this ambitious decarbonisation objective while at the same time ensuring security of energy supply and competitiveness. Finally, on March 2013, the EC adopted a Green Paper on 'A 2030 framework for climate and energy policies' (EC, 2013) with the aim, after a period of public consultation, of developing a 2030 framework for EU climate change and energy policies necessary to provide certainty and reduced regulatory risk for investors and to mobilise the funding needed; to support progress towards a competitive economy and a secure energy system; and to establish the EU's 2030 ambition level for GHG reductions in view of a new international agreement on climate change foreseen for 2015.

2.4.2 Renewable Energy Policy

The existing EU energy and environmental policy package includes numerous strategies, directives and regulations (Kanellakis et al., 2013). The most significant of these is the directive 2009/28/EC focussing on the promotion of the use of energy from renewable energy sources (RES). The directive 2009/28/EC approved mandatory targets of a 20% share of energy from RES in overall EU energy consumption by 2020 and a 10% target for each Member State regarding the share of RES consumption in transport by 2020. The directive translates the EU 20% target into individual targets for each Member State (Table 2.2).

The directive 2009/28/EC requires that transmission system operators and distribution system operators all over the EU guarantee the transmission and distribution of electricity produced from RES. When dispatching electricity, transmission system operators are to give priority to installations using RES so far as the secure operation of the national electricity system allows it. Additionally, steps towards the development of infrastructure for energy transmission and distribution, smart grids and storage facilities must be taken, so as to facilitate a secure operation of the electricity system as it integrates electricity production from RES. In addition, the directive establishes cooperation mechanisms by which Member States can join together to develop the RES. Using such mechanisms aims at overcoming national approaches towards a joint European perspective to the development of renewable energy. They include:

- Statistical transfers whereby one Member State with a surplus of renewable energy can sell it statistically to another Member State, whose RES may be more expensive. One Member State gains a revenue, at least covering the cost of developing the energy; the other gains a contribution towards its target at lower cost.
- Joint projects whereby a new renewable energy project in one Member State can be cofinanced by another Member State and the production shared statistically between the two. Joint projects can also occur between a Member State and a third country, if the electricity produced is imported into the EU (eg, from North Africa).
- Joint support schemes whereby two or more Member States agree to harmonise all or part of their support schemes for developing renewable energy, to clearly integrate the energy into the single market and share out the production according to a rule.

The EC estimates Member States' progress in the promotion and use of RES along the trajectory towards the 2020 targets (EREC, 2013). As seen in Table 2.2, most Member States have experienced significant growth in renewable

TABLE 2.2 Overview of Member States' Renewable Progress
and Target – Percent (%) (EC, 2013)

Member State	Renewable Energy Source (RES) Share in Gross Final Energy Consumption		RES Target Established by the 2009/28/EC Directive	National Targets Established by NREAPs
	2005	2013		
Austria	23.8	29.6	34.0	34.2
Belgium	2.3	6.2	13.0	13.0
Bulgaria	9.2	10.2	16.0	18.8
Cyprus	2.6	2.8	13.0	13.0
Czech Republic	6.1	8.5	13.0	13.5
Denmark	16.0	24.2	30.0	30.5
Estonia	17.5	12.7	25.0	25.0
Finland	28.6	29.2	38.0	38.0
France	9.5	9.0	23.0	23.3
Germany	6.0	10.3	18.0	19.8
Greece	7.2	10.7	18.0	20.2
Hungary	4.5	8.3	13.0	14.7
Ireland	2.8	6.2	16.0	16.0
Italy	5.1	16.5	17.0	16.2
Latvia	32.3	36.1	40.0	40.0
Lithuania	17.0	18.1	23.0	24.2
Luxembourg	1.4	3.6	11.0	8.9
Malta	0.0	1.5	10.0	10.2
Netherlands	2.1	4.2	14.0	14.5
Poland	7.0	8.7	15.0	15.5
Portugal	19.8	23.5	31.0	31.0
Romania	17.6	17.2	24.0	24.0
Slovakia	6.6	8.2	14.0	15.3
Slovenia	16.0	16.5	25.0	25.3

Continued

TABLE 2.2 Overview of Member States' Renewable Progress and Target – Percent (%) (EC, 2013) — cont'd

Member State	Renewable Energy Source (RES) Share in Gross Final Energy Consumption		RES Target Established by the 2009/28/EC Directive	National Targets Established by NREAPs
	2005	2013		
Spain	8.4	14.7	20.0	22.7
Sweden	40.4	34.8	49.0	50.2
United Kingdom	1.4	5.0	15.0	15.0
EU – 27	8.5	12.8	20.0	–

NREAPs, National renewable energy action plans.

energy. Austria, Finland, Latvia and Sweden have the highest share of renewables in their energy mix, between 30% and 45%; Denmark, Estonia, Lithuania, Portugal and Romania show the penetration of renewables in total energy mix between 20% and 25%.

In spite of fact, that Member States have experienced growth in RES, based on a variety of studies (EREC, 2013) and research on evaluating EU renewable energy policy (Jacobsson et al., 2009; Klessmann et al., 2011; Kitzing et al., 2012) the achievement of RES targets is still hindered by numerous barriers:

- Economic and market barriers. RES are not cost-competitive under current market conditions: high capital costs, unfavourable market pricing rules, subsidies for competing fuels, long reinvestment cycles of building-integrated technologies, etc.; limited access to finance and high cost of capital due to high perceived risk; power markets that are not prepared for RES; lack of access to the power markets, exercise of market power by large players, design not favourable for supply-driven RES, etc.
- Legal barriers, including inefficient administrative procedures (high number of authorities involved, lack of coordination among authorities, lack of transparent procedures, long lead times, high costs for applicants, etc.); RES not or insufficiently considered in spatial planning; no or insufficient standards and codes for RES equipment (specifications not well defined, not expressed in EU and international standards, etc.); tenancy law and ownership law impede the development of building-integrated RES technologies.
- Infrastructural and grid-related barriers (mainly power grids, but also gas and district heat). Grid access is difficult to obtain (transmission system operator and distribution network operator not open to RES, lack of transparent procedures, long approval times, unfavourable cost allocation

leading to high grid connection costs); lack of available grid capacity (weak grid environment, lack of interconnection capacity, no or slow grid reinforcement and/or extension, grid congestion leading to curtailment).

- Supply chain bottlenecks. Restricted access to technologies (only a few technology providers, lack of production capacity, a lack of R&D capacity); bottlenecks regarding feedstock supply (eg, steel, silicon, etc.).
- Information and social acceptance barriers. Lack of knowledge (about benefits of RES, about available support measures, etc.); lack of acceptance ('not in my back yard' opposition to RES plants and power lines, public concerns about sustainability of biofuels, etc.).
- Economic instability. The changed economic climate has clearly an impact (fully or partially) on the development of new RES projects and national renewable energy action plans commitments.

To overcome the these barriers, the EC established a series of tools aiming at supporting RES diffusion. Among them, a relevant role is played by the European Technology Platforms concerning wind energy, photovoltaic technology, biofuels technology, electricity networks, renewable heating and cooling, zero emission fossil fuel power plants, sustainable nuclear technology and the fuel cells and hydrogen. These platforms constitute as many supporting tools, whose mission is to develop a strategy and corresponding implementation plan for research and technology development, innovation and market deployment of renewable energy.

2.4.3 Energy Efficiency Policy

Aside from renewable energy target, the EC has committed to reach the objective of 20% primary energy savings in 2020 compared to a baseline. To meet these objectives, the EU has realised a variety of energy efficiency policies and directives. The most important initiatives relevant for energy efficiency include the Effort Sharing Decision, the Energy Performance of Buildings Directive, the Eco-design Directive, the Labelling Directive, minimum standards for electric motors, minimum standards for commercial lighting, the Energy Efficiency Plan and the Energy Efficiency Directive.

The Energy Efficiency Directive includes general energy efficiency policies and measures addressing specific energy consumption sectors as, eg, buildings, transport, industry, energy audits and management systems. The Energy Efficiency Directive also establishes a requirement for Member States to set indicative national energy efficiency targets for 2020, which are to be determined nationally and can be based on different indicators such as final energy consumption, final energy savings or energy intensity as shown in Table 2.3.

In view of energy efficiency progress, the European Environment Agency (EEA) conducted an evaluation of progress on energy efficiency (EEA, 2013), concluding in overall terms that progress in energy efficiency in the last years across Member States was rather modest. As seen in Table 2.4, only four

TABLE 2.3 Overview of Member States' National Energy Efficiency Targets (EC, 2013)

Member State	Indicative National Energy Efficiency Targets for 2020	Absolute 2020 Level of Energy Use (Mtoe)	
		Primary	Final
Austria	Final energy consumption of 1100 PJ	31.5	26.3
Belgium	Reducing primary energy consumption by 18% compared to projections for 2020	43.7	32.5
Bulgaria	Increase of energy efficiency by 25% until 2020 (5 Mtoe primary energy savings in 2020) and 50% energy intensity reduction by 2020 compared to 2005 levels	15.8	9.16
Cyprus	0.463 Mtoe energy savings in 2020 (14.4% reduction compared to BAU)	2.8	2.2
Czech Republic	Energy use shall be 20% more efficient by 2020	39.6	24.4
Denmark	Primary energy consumption of 744.4 PJ (17.781 Mtoe) in 2020	17.8	14.8
Estonia	Stabilisation of final energy consumption in 2020 at the level of 2010	6.5	2.8
Finland	310 TWh of final energy consumption in 2020	35.9	26.7
France	17.4% reduction of final energy consumption in 2020 compared to a baseline	236.3	131.4
Germany	Annual improvement of energy intensity by 2.1% pa. on average until 2020	276.6	194.3
Greece	Final energy consumption level of 20.5 Mtoe	27.1	20.5
Hungary	1113 PJ primary energy consumption in 2020 (236 PJ) savings compared to business-as-usual), resulting in 760 PJ final energy consumption	26.6	18.2
Ireland	20% energy savings in 2020 along with a public sector energy-saving target of 33%	13.9	11.7

Italy	20 Mtoe primary energy reduction by 2020; 15 Mtoe final energy reduction by 2020	158.0	126.0
Latvia	Primary energy savings in 2020 of 0.670 Mtoe (28 PJ)	5.37	4.47
Lithuania	17% reduction in final energy use compared to 2009 level (reduction of 740 ktoe)	–	5.4
Luxembourg	Energy use shall be 20% more efficient by 2020	4.48	4.24
Malta	22% energy or 237.019 toe savings target by 2020	0.825	0.493
Netherlands	1.5% energy savings per year (partial)	60.7	52.1
Poland	13.6 Mtoe primary energy savings in 2020	96.4	70.4
Portugal	Reduction of primary energy use in 2002 by 25% compared to projections	22.5	17.4
Romania	Reduction of 10 Mtoe (19%) in the primary energy consumption	42.99	30.32
Slovakia	3.12 Mtoe of final energy savings for the period 2014–2020	16.2	10.4
Slovenia	10.809 GWh energy savings by 2020	–	–
Spain	20% energy savings to be achieved by 2020	121.6	82.9
Sweden	Energy use shall be 20% more efficient by 2020 compared with 2008 and a 20% reduction in energy intensity between 2008 and 2020	45.9	30.3
United Kingdom	Final energy consumption in 2020 of 129.2 Mtoe on a net calorific value basis	177.6	157.8

BAU, Business as usual.

TABLE 2.4 Overview of Member States' Energy Efficiency Progress (EEA, 2013)

Member State	Primary Energy Consumption (Mtoe)		Absolute Change Primary Energy Consumption Index, 2005=100 (%)	Final Energy Consumption (Mtoe)		Absolute Change Final Energy Consumption Index, 2005=100 (%)
	2005	2011		2005	2011	
Austria	32.7	32.4	99.0	28.1	27.3	97.1
Belgium	51.5	52.0	101.1	36.6	38.9	106.3
Bulgaria	19.2	18.8	97.7	10.0	9.3	92.6
Cyprus	2.4	2.6	106.5	1.8	1.9	104.4
Czech Republic	42.3	40.7	96.3	26.0	24.6	94.7
Denmark	19.5	18.7	96.0	15.5	14.7	94.7
Estonia	5.4	6.1	113.8	2.9	2.8	99.2
Finland	33.7	34.4	102.0	25.5	25.2	98.7
France	260.3	245.4	94.3	162.4	148.1	91.2
Germany	314.7	286.4	91.0	229.5	207.1	90.2
Greece	30.6	27.0	88.3	20.8	18.8	90.5
Hungary	25.5	23.3	91.0	18.2	16.3	89.6
Ireland	14.8	13.6	91.7	12.5	10.8	86.3
Italy	179.9	161.9	90.0	134.6	122.3	90.9
Latvia	4.4	4.1	94.5	4.0	4.0	99.0

Lithuania	8.0	5.8	73.3	4.6	4.7	102.1
Luxembourg	4.8	4.6	95.2	4.4	4.3	96.3
Malta	0.9	1.1	118.1	0.4	0.4	114.7
Netherlands	69.5	67.4	96.9	52.3	50.7	96.9
Poland	88.5	97.3	109.9	58.2	64.7	111.2
Portugal	24.9	22.2	89.1	19.0	17.4	91.5
Romania	36.8	33.9	92.4	25.1	22.6	90.0
Slovakia	17.6	16.0	91.0	11.1	10.8	97.5
Slovenia	7.0	7.1	102.2	4.9	5.0	101.6
Spain	136.0	121.8	89.5	97.5	86.5	88.8
Sweden	49.4	47.6	96.2	33.6	32.2	95.7
United Kingdom	222.5	190.7	85.7	152.3	132.0	86.7
EU – 27	1711.0	1591.0	93.0	1198.2	1109.4	92.6

Member States (Bulgaria, Denmark, France and Germany) are making good progress in reducing energy consumption and primary energy intensity. The other Member States, however, did not make significant energy efficiency progress; the current energy efficiency policies are not sufficiently developed and implemented.

Financing energy efficiency projects seems to be the highest barrier (EEA, 2013). According to EEA, to meet energy efficiency targets it will require €900 billion over the period 2010–20. The uncertainty range is from €800 to €1200 billion. The breakdown of the €900 billion is as follows: buildings, €400 billion (uncertainty range €350–650 billion); transportation, €400 billion (uncertainty range €300–500 billion); industry, €100 billion.

To overcome the financial barrier, the EC has introduced some specific supporting tools. For instance, the Global Energy Efficiency and Renewable Energy Fund (GEEREF) aims at mobilising private investments in energy efficiency and renewable energy projects. It establishes a public–private partnership by offering ways of risk sharing and cofinancing for projects investing in energy efficiency and renewable energy. GEEREF participation ranges from between 25% and 50% for medium to high-risk operations to 15% for low-risk operations. Provision is also made for dedicated technical assistance funds.

2.4.4 Market Integration

The internal energy market has been at the centre of the EU energy initiatives. The EC has established a series of directives for electricity and gas regulations, Directive concerning common rules for the internal market in electricity and Directive concerning common rules for the internal market in natural gas. These directives require designation of independent regulatory authorities for all Member States that must be legally distinct and functionally independent from any other public or private entity. The duties of these regulatory authorities are to oversee and monitor the whole electricity and gas market, facilitating their regular functioning and the rights and obligations of each of the legal entities and undertakings involved in the markets.

Within the context of these directives, Member States shall establish transmission system operators. Transmission system operators shall build sufficient cross-border capacity to integrate the European transmission infrastructure. Every year, they shall submit to the regulatory authority a 10-year network development plan indicating the main infrastructure that needs to be built or modernised as well as the investments to be executed over the next 10 years.

Additionally, Member States shall designate distribution system operators. Distribution system operators are mainly responsible for ensuring the long-term capacity of the system in terms of the distribution of electricity and gas, operation, maintenance, development and environmental protection; ensuring transparency with respect to system users; providing system users with information; covering energy losses and maintaining reserve capacity.

However, in spite of existing supporting policy and the fact that the EU Member States had endorsed 2014 as the target date for completing the EU's internal energy market (for this purposes, EC has also presented a Communication which gives guidance to Member States on how to make the most of public interventions, how to reform existing ones – especially renewable energy subsidy schemes – and how to effectively design new ones [EC, 2013]), the deadline has passed and there is a deep concern that the goal of creating an integrated wholesale energy market is still some ways off. Based on studies assessing the EU policies addressing market integration (Brunekreeft et al., 2012) the prospect of a single market for energy will take even longer to achieve owing numerous barriers:

- One of the biggest problems is the lack of investments. Achieving an integrated energy market will require enormous investment in new infrastructure, especially if the EU needs to be able to increase the share of energy from renewable sources to the levels that are predicted. New flexible grids will need to be built to bring power from offshore wind farms to major population centres. More interconnectors will be needed to ensure that gas and electricity can flow freely within the EU's internal market.
- One of the problems is the strong position of local and regional suppliers. Market entry is also made less attractive for new and foreign companies because of the already strong position of established local and regional suppliers. On the one hand, some of these suppliers are subject to strong political influence and on the other hand they are able to establish a good position thanks to their regional proximity to their customers.
- One of the problems is renewable energy integration in European electricity grids: geographical distribution (the geographical distribution of renewable energy on the one hand and demand on the other hand often do not coincide, nor are renewables necessarily located close to current generation centres – this is especially true for offshore wind parks and concentrated solar power deployed in desert areas); variability and intermittency (renewable power generation like wind and solar power is that it can be interrupted, and this variability affects the stability of the power produced); grid access is also difficult to obtain (transmission system operator and distribution network operator still not open to RES); and a lack of available grid capacity (weak grid environment, lack of interconnection capacity).

As a supporting tool to overcome all barriers the EC established the Agency for the Cooperation of Energy Regulators (ACER) and the European Network of Transmission System Operators, with the purpose of assisting regulatory authorities in exercising regulatory tasks performed in the Member States and, where necessary, to coordinate their action. ACER is responsible for adopting, under certain conditions, individual decisions on technical issues. It may make recommendations with the aim of promoting the exchange of good practice between regulatory authorities and market players. It shall also provide a framework for cooperation between the national regulatory authorities.

2.4.5 Energy Networks

Energy infrastructure is a top priority of the EU energy policy strategies; electricity transmission, gas and oil pipelines, smart grids, storage of energy and CO_2 transport are essential elements of EU present and future energy systems. Therefore, the EC has adopted the Energy Infrastructure Package for 2020 and beyond.

As seen in Fig. 2.2, the Energy Infrastructure Package clearly identifies 12 priority corridors and areas where accelerated investments are needed to finalise

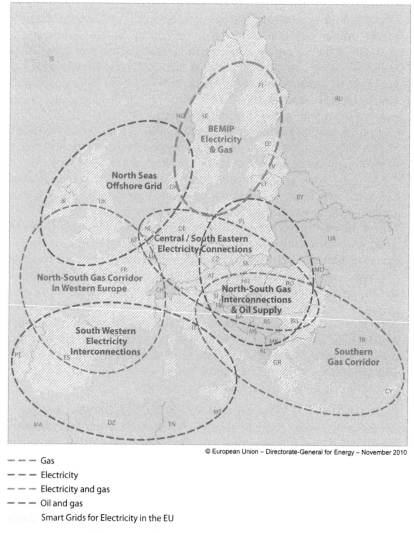

© European Union – Directorate-General for Energy – November 2010

− − − Gas
− − − Electricity
− − − Electricity and gas
− − − Oil and gas
          Smart Grids for Electricity in the EU

FIGURE 2.2   European priority corridors for electricity, gas and oil (EC, 2010).

the single energy market. These priorities cover different geographic regions or thematic areas in the field of electricity transmission and storage, gas transmission, storage and liquefied or compressed natural gas infrastructure, carbon dioxide transport and oil infrastructure (eg, development of the North Sea offshore grid, North–South electricity interconnections in Central Eastern and South Eastern Europe, Southern Gas Corridor, oil supply connections in Central Eastern Europe).

To find the best value-added investments, the EC selects the projects of common interests (PCIs) and these selected projects will have an accelerated permitting procedure and financial assistance from EU funds. The concrete measures proposed by the EC provide appropriate answers to the infrastructure challenge of Europe by mitigating several types of risks associated with energy infrastructure development. A dedicated budget provided by the EU has been distributed between a limited number of projects with common interests, resulting in a significantly lower capital expenditure of the selected investments. Additionally, the acceleration of the permitting procedure reduces the risk of delays in the phase of project execution. Furthermore, the establishment of regional groups enhances regional cooperation among EU Member States.

2.4.6 Security of Energy Supply

In order to deal with energy security challenges, the EC has established the extensive developments of the concept toward energy security mainly focussing on maintaining good relations with oil and gas producer countries as well as on prevention and crisis management, as a response to operational or technical disruptions and natural calamities, and to disruptions caused by temporary political conflicts.

The EC has adopted a special directive imposing obligations on Member States aimed at ensuring a high level of security of oil supply through reliable and transparent mechanisms based on solidarity amongst Member States, maintaining minimum stocks of crude oil and petroleum products and putting in place the necessary procedural means to deal with a serious shortage.

Member States shall ensure that emergency stocks and specific stocks are available and physically accessible for the purposes of this Directive. They shall establish arrangements for the identification, accounting and control of those stocks so as to allow them to be verified at any time. This requirement also applies to any emergency stocks and specific stocks that are commingled with other stocks held by economic operators.

In the context of external relations, the EC has established a wide range of initiatives and strategic dialogues with Russia, Central Asia, Caspian Sea, the Mediterranean Sea, the Arab states on the Persian Gulf, and Turkey. All these initiatives call for a new type of partnership in regulatory cooperation, energy security and safety, market access and investment protection, using appropriate cooperation instruments and agreements. Such partnerships should also extend to the efficient use of available resources, as well as joint assessment of long-term energy supply and demand perspectives.

2.4.7 Energy Technology and Innovation Policy

To provide technological support for low-carbon transition, the energy research is implemented through the EU's general framework programme for research such as the European Strategic Energy Technology (SET) plan. The SET plan includes such innovative group as the European Wind Initiative (focus on large turbines and large systems validation and demonstration relevant to on- and off-shore applications); Solar Europe Initiative (focus on large-scale demonstration for photovoltaic and concentrated solar power); Bioenergy Europe Initiative (focus on next-generation biofuels within the context of an overall bioenergy use strategy); European CO_2 capture, transport and storage initiative (focus on the whole system requirements, including efficiency, safety and public acceptance); European electricity grid initiative (focus on the development of the smart electricity system); Sustainable nuclear fission initiative (focus on the development of new-generation reactors technologies); the European Research Area (ERA-Net) scheme (encouraging Member States to coordinate research and development programmes); the Joint Technology Initiatives and Smart Cities and Communities programme.

The EC has also established Horizon 2020, the EU's new funding programme for research and innovation. Horizon 2020 brings together higher education institutions, research centres and businesses, all existing EU research and innovation funding, including the Framework Program for Research, the innovation-related activities of the Competitiveness and Innovation Framework Program and the European Institute of Innovation and Technology.

However, both the SET plan as well as Horizon 2020 is at risk of failing to deliver its objectives owing to inadequacy of finance and uncertainty about the future policy framework. While the EU level of investment in low-carbon energy in 2011 was 94 billion USD, comparing favourably to the amount of 50 billion USD invested by the United States and China, the EU level of investment was expected to be lower in future, owing to economic recession in some EU Member States (Eurelectric, 2013). Economic and investment uncertainties could impact on EU leadership in energy innovative sectors in the global market. China has already developed a strong base in the manufacture of renewable energy equipment, notably for the solar and wind industry. Therefore, to deal with inadequate funding of EU innovative programmes, the EC must revise both the SET plan and Horizon 2020 with a view of how these innovative programmes will be financed.

2.5 CONCLUSIONS

The present chapter has focused on the patterns of energy production and consumption in the EU, insisting on the existing unbalances in energy resources and energy demand and unfolding the challenges the latter represent in relation to European energy security. Moreover, it has sketched out the historical background for the development and consolidation of a common EU energy policy, highlighting its main fields of actions and features, and reflecting upon its potentials and pitfalls.

If one main message is to be distilled from the information presented in the various sections, it is that there are at least three keywords and an alert that can describe the current situation of the EU and its Member States in the field of energy security and low-carbon transition strategies. The keywords are dependency, consumption and integration. The alert refers to the current economic crisis.

Dependency is a major keyword to describe the EU energy scenario because both in terms of trends (ie, all those market, societal, economic and geopolitical features) and of strategies (ie, the directions to take in order to achieve expected results) it seems that on one hand the EU is and will be depending from imports for the vast majority of its energy needs (with an increasing dependency rate of 9% from 1999 to 2009), while on the other hand the possibility that a fair degree of energy independence could be achieved rely both on much higher investments on renewables and on the use of less sustainable energy sources (as in the case of nuclear energy, coal and oil). To this extent, the whole of the EU seems to be moving away both from fossil fuels (−5.9% in 10 years since 2000) and from nuclear energy (a more modest 0.7%), but the situation varies dramatically in the different Member States, with only Latvia and Sweden being the exceptions that can account to renewables for one-third of their gross inland primary consumption. In addition, there is an internal issue of dependency within the EU because Member States act differently on the market (there is not anything like a 'single EU buyer'), have different national and internal resources (as in the case of the oil available to countries facing the North Sea), have different policies regarding the balance between traditional sources and RES notwithstanding the role of EU addresses and directives, and have to face market and societal inertia that may hamper efforts towards change.

Consumption seems like the other face of the coin, at least because it seems to exacerbate the dependency of the EU (and of some Member States more than others) from a certain level of provision that may require increasing expenditures (as in the case of market fluctuations) or complex energy mixes (as in the case of those countries that have to rely almost totally on energy import). The EU, in fact, according to the latest available data (2010) was the third major global consumer (13.4% of the total world consumption, behind China and the United States), but what is more worrying is that different economic sectors rely on different energy sources and that each sector depends heavily on one or few energy sources: that is almost the contrary of what should be expected in terms of consumption to avoid an excessive dependency rate. Again, also in this case there is a consumption issue at the EU level but national situations may vary consistently. To be more specific, consumption calls for a closer look at lifestyles, societal organisation in the energy field, different patterns according to local environmental and cultural conditions, etc. At the local level, for instance, it is possible to see the perverse effects of a missing consumption policy in terms of industrial competitiveness and the related phenomenon of the 'carbon leakage' – that is, relocalisation because of increased production costs in the EU.

Integration calls for a better harmonisation of policies and interventions at different levels, from the local to the EU and beyond. It seems, in fact, that the issues that have been raised in terms of dependency and consumption would greatly benefit from a coherent strategy able to develop into coherent policies in the whole of the EU. Differentiation of energy sources and energy imports and the promotion of self-reliance are at the core of EU energy strategies but the filtering down of directives to Member States and regions seems to weaken the effort. For instance, the need for a single, coherent energy market has been made explicit but still the process is in the making. A second dimension of integration is related, of course, to the energy mix that – in order to be part of a successful transition towards low-carbon societies – needs to be oriented towards a more sustainable use of traditional energy sources and to an increasing role of renewables.

The alert, finally, is on the *crisis* in terms of an effective analysis of the current situation. Data are, in fact, somehow misrepresented because of (1) their availability, which often is up to the first years of the crisis and thus not fully revealing the impact, and (2) the possibility that such a deep and long crisis will permanently produce changes in the way single citizens and the whole society relate to energy consumption.

REFERENCES

Averre, D., 2010. The EU, Russia and the shared neighbourhood: security, governance and energy. European Security 19 (4), 531–534.

Bahgat, G., 2011. Energy Security: An Interdisciplinary Approach. Wiley.

Bilgin, M., 2009. Geopolitics of European natural gas demand: supplies from Russia, Caspian and the Middle East. Energy Policy 37 (11), 4482–4492.

Birchfield, V.L., Duffield, J.S., 2011. Toward a Common European Union Energy Policy. Problems, Progress and Prospects. Palgrave MacMillan, New York.

Brunekreeft, G., Brandstätt, C., Friedrichsen, N., Meyer, R., Meyer, S., Palovic, M., 2012. European Internal Electricity Market for Consumers - Opportunities and Barriers to Cross-Border Trade between Germany and Austria. Bremer Energy Institute.

Bunse, K., Vodicka, M., Schönsleben, P., Brülhart, M., Ernst, F.O., 2011. Integrating energy efficiency performance in production management – gap analysis between industrial needs and scientific literature. Journal of Cleaner Production 19 (6), 667–679.

Chakarova, V., 2010. Oil Crises and International Cooperation: Do International Institutions Matter? The 7th SGIR Pan-European Conference. Stockholm.

Chester, L., 2010. Conceptualising energy security and making explicit its polysemic nature. Energy Policy 38 (2), 887–895.

Clastres, C., 2011. Smart grids: another step towards competition, energy security and climate change objectives. Energy Policy 39 (9), 5399–5408.

Dittmar, M., 2012. Nuclear energy: status and future limitations. Energy 37 (1), 35–40.

EC, 2000. Towards a European Strategy for the Security of Energy Supply. COM (2000) 769 final report.

EC, 2007. Strategic Energy Technology Plan (SET-Plan): Towards a Low Carbon Future. COM (2007) 723 final report.

EC, 2008. European Energy and Transport. Trends to 2030-Update 2007. Publications Office of the European Communities, Luxembourg.

EC, 2010. Communication from the Commission to the European Parliament, the Council the European Economic and Social Committee and the Committee of the Regions, Energy Infrastructure Priorities for 2020 and Beyond. COM (2010) 677 final report.

EC, 2011. EU-Russia Energy Dialogue. The First Ten Years: 2000–2010. (Brussels, EU).

EC, 2012. European Commission, Communication from the Commission to the European Parliament, the Council, the European Economic and Social Committee and the Committee of the Regions, Report from the commission to the European parliament and the council, progress towards achieving the Kyoto objectives.

EC, 2013. European Commission, Communication from the Commission to the European Parliament and the Council. Renewable Energy: Progressing Towards the 2020 Target. (Brussels).

EEA, 2013. Trends and Projections in Europe. Tracking Progress Towards Europe's Climate and Energy Targets Until 2020.

EEC, 2013. Smarter, Greener, More Inclusive? Indicators to Support the Europe 2020 Strategy. Publications Office of the European Union, Luxembourg.

EREC, 2013. European Renewable Energy Council. EU Tracking Roadmap: Keeping Track of Renewable Energy Targets Towards 2020.

Eurelectric, 2013. Power Choices Pathways to Carbon-neutral Electricity in Europe by 2050. (Brussels).

Faruqui, A., Harris, D., Hledik, R., 2010. Unlocking the €53 billion savings from smart meters in the EU: how increasing the adoption of dynamic tariffs could make or break the EU's smart grid investment. Energy Policy 38 (10), 6222–6231.

Finon, D., 2011. The EU foreign gas policy of transit corridors: autopsy of the stillborn Nabucco project. Energy Review 35 (1), 47–69.

Fisher, S., 2012. Carbon Capture and Storage: The Europeanization of a Technology in Europe's Energy Policy? European Energy Policy. An Environmental Approach. Edward Elgar, Cheltenham.

Glachant, J.-M., Lévêque, F., 2009. The Electricity Internal Market in the European Union: What to Do Next? Electricity Reform in Europe. Towards a Single Energy Market. Edward Elgar, Cheltenham.

Groenenberg, H., Ferioli, F., Heuvel, S., Kok, M., Manders, T., Slingerland, S., Wetzelaer, B.J.H.W., 2008. Climate, Energy Security and Innovation; an Assessment of EU Energy Policy Objectives. ECN, Petten.

IEA, 2007. Energy Security and Climate Policy. Assessing Interaction, Paris.

IPCC, 1990. Climate Change. The IPCC Scientific Assessment. Cambridge University Press.

Jacobsson, S., Bergek, A., Finon, D., Lauber, V., Mitchell, C., Toke, D., Verbruggen, A., 2009. EU renewable energy support policy: faith or facts? Energy Policy 37 (6), 2143–2146.

Kanellakis, M., Martinopoulos, G., Zachariadis, T., 2013. European energy policy. A review. Energy Policy 62, 1020–1030.

Kemp, R., Pontoglio, S., 2011. The innovation effects of environmental policy instruments. A typical case of the blind men and the elephant? Ecological Economics 72, 28–36.

Kitzing, L., Mitchell, C., Morthorst, P., 2012. Renewable energy policies in Europe: converging or diverging? Energy Policy 51, 192–201.

Klessmann, C., Held, A., Rathmann, M., Ragwitz, M., 2011. Status and perspectives of renewable energy policy and deployment in the European Union – what is needed to reach the 2020 targets? Energy Policy 39 (12), 7637–7657.

Kolk, A., 2012. The role of consumers in EU energy policy. Carbon Management 3 (2), 175–183.

Nilsson, M., Rickne, A., 2012. Governing Innovation for Sustainable Technology: Introduction and Conceptual Basis. Paving the Road to Sustainable Transport. Governance and Innovation in Low-Carbon Vehicles, Routledge, Oxford.

Quemada, J.M.M., García-Verdugo, J., Escribano, G., 2011. Energy Security for the EU in the 21st Century. Markets, Geopolitics and Corridors, Abingdon, Routledge.

Rio, P., 2012. Analysing future trends of renewable electricity in the EU in a low-carbon context. Renewable and Sustainable Energy Reviews 15 (5), 2520–2533.

Roos, I., Soosaar, S., Volkova, A., Streimikene, D., 2012. Greenhouse gas emission reduction perspectives in the Baltic States in frames of EU energy and climate policy. Renewable and Sustainable Energy Reviews 16 (4), 2133–2146.

Schmidt, T.S., Schneider, M., Rogge, K.S., Schuetz, M.J.A., Hoffmann, V.H., 2012. The effects of climate policy on the rate and direction of innovation: a survey of the EU ETS and the electricity sector. Environmental Innovation and Societal Transitions 2, 23–48.

Timilsina, G.R., Kurdgelashvili, L., Narbel, P.A., 2012. Solar energy: markets, economics and policies. Renewable and Sustainable Energy Reviews 16 (1), 449–465.

Vinois, J.A., 2010. European Energy Policy – An Integrated Approach on Energy Security. JRC-Institute for Energy.

Visschers, V.H.M., Keller, C., Siegrist, M., 2011. Climate change benefits and energy supply benefits as determinants of acceptance of nuclear power stations: investigating an explanatory model. Energy Policy 39 (6), 3621–3629.

Wüstenhagen, R., Menichetti, E., 2012. Strategic choices for renewable energy investment: conceptual framework and opportunities for further research. Energy Policy 40, 1–10.

Young, R., 2011. Foreign Policy and Energy Security: Markets, Pipelines and Politics. Toward a Common European Union Energy Policy. Problems, Progress and Prospects. Palgrave MacMillan, New York.

Chapter 3

A Study of Russia as Key Natural Gas Supplier to Europe in Terms of Security of Supply and Market Power

A. Prahl[1], K. Weingartner[2]

[1]Ecologic Institute, Berlin, Germany; [2]Ecologic Institute US, Washington, DC, United States, formerly Ecologic Institute, Berlin, Germany

3.1 INTRODUCTION

Natural gas is a significant part of the European Union's (EU) energy mix and the ability to assure its secure supply to meet energy needs is an important component of any future EU energy security strategy. Although natural gas consumption has been declining in the EU, this decline is not keeping pace with EU domestic production declines. Recognising the import dependency that this creates for the EU, the role of key natural gas exporters to the EU must be a prominent consideration when discussing EU energy security. Russia's position as the main exporter of natural gas to the EU calls into question the impacts of this relationship in terms of the security of supply and market power. Meanwhile the EU is implementing policies to reduce its exposure to Russian gas, with implications to the Russian market in terms of the security of demand. This chapter takes an in-depth look at the current realities of the natural gas market in the EU and Russia, their relationship and how both the EU and Russia are taking measures to promote their own energy security, albeit through different approaches. Extrapolating into the future, the chapter comments on what the role of natural gas could be in the EU in the near- and midterm and how this affects EU energy security and the relationship with Russia.

3.2 CURRENT LOOK AT NATURAL GAS IN THE EU

3.2.1 EU Natural Gas Consumption Over the Peak?

Domestic natural gas demand has been on the decline recently. Though natural gas consumption varies across Member States, from 2004 to 2014 EU natural

Low-carbon Energy Security from a European Perspective. http://dx.doi.org/10.1016/B978-0-12-802970-1.00003-6
Copyright © 2016 Elsevier Ltd. All rights reserved.

gas consumption fell by 21.2% from 491 bcm to 386.9 bcm. Over this same time period, all major natural gas consumers in Europe have seen decreases in consumption, mainly Germany (−17.5%), France (−20.4), the Netherlands (−21.5%), Italy (−23.1%) and the United Kingdom (−31.5%). In 2014, natural gas consumption represented 24% of total EU primary energy consumption that year (BP, 2015a).

Different causes have been attributed to the recent decline in natural gas consumption within the EU. From 2010 to 2014, renewable energy consumption in the EU increased by 31% (BP, 2015a). Coal power plants come before gas plants in the merit order and the low EU Emissions Trading System (EU ETS) price have kept coal an economic energy carrier. Furthermore, primary energy consumption in the EU decreased by 5.2% between 2010 and 2013(Eurostat, 2015c), which was due to various factors. Progress had been made in energy efficiency, with 6.7% less energy needed per unit of GDP from 2010 to 2013 (Eurostat, 2015b). Moreover, the warm winter in 2012, as well as the recession in 2012 and low GDP growth in 2013 and 2014 contributed to the reduction in primary energy consumption (Eurostat, 2015a).

Fig. 3.1 shows the development of the aggregated natural gas consumption and production of today's EU Member States from 1970 until 2014.

The majority of natural gas within the EU is currently consumed in the residential sector with approximately 27%, for heating purposes. Furthermore, 25% are used by the power industry to generate electricity. Also the industry and services sectors use substantial amounts of gas, with 23% and 12% of overall consumption (Eurostat statistical books, 2015). Fig. 3.2 shows the distribution of natural gas consumption across sectors within the EU.

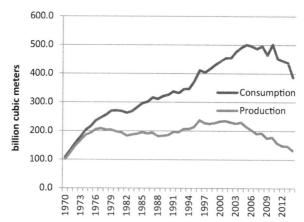

FIGURE 3.1 Natural gas consumption and production of today's European Union Member States from 1970 until 2014. *Data from: BP, 2015a. BP Statistical Review of World Energy June 2015. Retrieved from:* http://www.bp.com/content/dam/bp/pdf/Energy-economics/statistical-review-2015/bp-statistical-review-of-world-energy-2015-full-report.pdf *(excludes Estonia, Latvia and Lithuania prior to 1985 and Slovenia prior to 1991).*

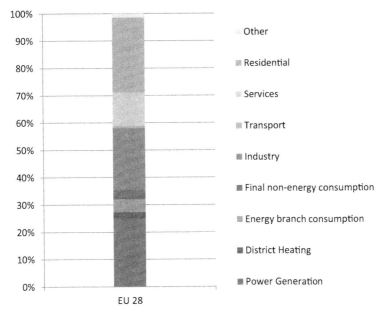

FIGURE 3.2   EU-28 natural gas consumption according to sectors in 2013. *Data from: Eurostat statistical books, 2015. Energy Balance Sheets 2013 Data (No. ISSN 1830–7558). Eurostat. Retrieved from:* http://ec.europa.eu/eurostat/documents/3217494/6898731/KS-EN-15-001-EN-N. pdf/e5851c73-9259-462e-befc-6d037dc8216a.

3.2.2 Dwindling Domestic Production

Whereas some decades ago, European gas demand was almost entirely met by European indigenous production (Dickel et al., 2014), after peaking in 1996, EU gas production remained constant until 2004 when it began to decline (Söderbergh et al., 2010). This decline continues, demonstrated by the fact that in 2014, EU gas production dropped by almost 10% in comparison to 2013, representing the lowest production level since the start of the 1970s. In 2014, indigenous conventional gas production in the EU was 132.3 bcm (BP, 2015a). Thus, 2014 natural gas production could theoretically cover 34% of the same years' consumption.

At the Member State level, natural gas production has declined almost across the board. Although the United Kingdom (+0.3%) and Romania (+5.4%) experienced production increases from 2013 to 2014, all other EU natural gas producers experienced production decreases. Production in the Netherlands fell by 18.7%, with Italy (−7.6%) and Germany (−6.1%) following similar patterns (BP, 2015a).

The largest natural gas producers within the EU are projected to show further decreases in production in the future. In 2014, the Netherlands was the largest producer of natural gas, accounting for 42% of EU natural gas production. However, concerns over earth tremors have led to production reductions of 80%

in the most at risk areas and caps on production of 42.5, 42.5 and 40 bcm for 2014, 2015 and 2016, respectively. Although production caps are set for review after 2016, annual production is estimated to remain at 40 bcm until 2020 with levels reaching 16 bcm by 2030 (Dickel et al., 2014) (Table 3.1).

The United Kingdom was the second largest producer of natural gas in the EU in 2014, representing 27.7% of EU natural gas production. However, natural gas production in the United Kingdom has been in continuous decline since peaking in 2000. Furthermore, in 2004 the United Kingdom moved from being a net gas exporter to a net gas importer. According to a 2014 research report by the UK Energy Research Centre, around half of UK natural gas consumption is now met through imports. Though further development of the United Kingdom Continental Shelf (UKCS) and shale gas development onshore could slow production decline, projections show further decline in natural gas production of the UKCS of 5% per year (Bradshaw et al., 2014).

In combination with reduced production, reserves are being depleted. The reserve–production ratio shows the number of years that remaining reserves would last according to a continuation of the previous year's production rate, calculated by dividing remaining reserves at the end of a year by that year's production. For Europe's top natural gas producer, the Netherlands, this ratio was 14.3 in 2014. In the United Kingdom, the 2014 reserve–production ratio was 6.6 (BP, 2015a).

The remaining EU Member States producing natural gas are doing so to a much lesser extent. Together, Romania, Germany, Italy, Denmark and Poland contribute to 26% of EU natural gas production. Among these smaller producers, production output is also expected to decrease, with reserve–production ratios below 10 (BP, 2015a).

Significant future increases in EU domestic production through shale gas are mainly seen as unlikely, largely due to limitations which are predicted to

TABLE 3.1 Changes in Production From Peak Production to 2014 for Major EU Natural Gas Suppliers

Country	Peak Production Year	Peak Production (bcm)	2014 Production (bcm)	% Change From Peak Production Year to 2014
Netherlands	1977	82.3	55.8	−32.2%
United Kingdom	2000	108.4	36.6	−66.2%

Data from: BP, 2015a. BP Statistical Review of World Energy June 2015. Retrieved from: http://www.bp.com/content/dam/bp/pdf/Energy-economics/statistical-review-2015/bp-statistical-review-of-world-energy-2015-full-report.pdf.

prevent noteworthy shale gas production development until 2020 and possibly even into the 2030s. Both the International Energy Agency (IEA) (in 2013) and BP (in 2014) estimate that by the early 2030s, Europe will be producing shale gas at levels below 30 bcm (Bradshaw et al., 2014).

3.2.3 Natural Gas Imports Cover the Majority of EU Supplies

Due to the considerable spread between EU natural gas consumption and domestic production levels over the past decades (see Fig. 3.1), the EU has been required to rely on natural gas imports to make up the difference.

Table 3.2 shows that over the last years overall EU pipeline import amounts have remained fairly constant around 250 bcm per year, with 248 in 2014. There is also the import of liquefied natural gas (LNG), but with 45 bcm in 2014 this plays a much lesser role in EU natural gas supply (see also Table 3.4) and currently mainly affects Spain and the United Kingdom. Altogether, gas imports in 2014 where at 293 bcm, which amounts to some 75% of EU consumption (BP, 2015a).

In 2014, 41% of EU natural gas pipeline imports from outside EU territory came from Norway, making it the second largest gas supplier with 101 bcm coming into the EU in that year. The reserve–production ratio in 2014 for Norway was 17.7 (BP, 2015a) hinting towards stable availability in the medium term future. Estimates from Norway's Ministry of Petroleum project gas sales of 100–125 bcm for 2020, with decreases to 75–115 bcm in 2025 (Dickel et al., 2014). It is possible that the discovery of new reserves will change these estimates, but future extraction in the north is projected to be expensive (Dickel et al., 2014).

TABLE 3.2 Recent Development of Natural Gas Pipeline Imports From Russia and Norway

bcm	2004	2006	2008	2010	2012	2014
Russia	132	129	129	111	106	121
Norway	75	84	93	96	107	101
Other	34	51	9	51	39	26
Total	**241**	**264**	**231**	**258**	**252**	**248**

BP, 2005. BP Statistical Review of World Energy June 2005. BP. Retrieved from: http://www.bp.com/statisticalreview; BP, 2007. BP Statistical Review of World Energy June 2007. BP. Retrieved from: http://www.bp.com/statisticalreview; BP, 2009. BP Statistical Review of World Energy June 2009. BP. Retrieved from:http://www.bp.com/statisticalreview; BP, 2011. BP Statistical Review of World Energy June 2011. BP. Retrieved from: http://www.bp.com/statisticalreview; BP, 2013. BP Statistical Review of World Energy June 2013. BP. Retrieved from: http://www.bp.com/statisticalreview; BP, 2015a. BP Statistical Review of World Energy June 2015. Retrieved from: http://www.bp.com/content/dam/bp/pdf/Energy-economics/statistical-review-2015/bp-statistical-review-of-world-energy-2015-full-report.pdf.

The largest gas supplier to the EU is Russia. In 2014, it accounted for 49% of EU natural gas pipeline imports from outside EU territory or 121 bcm of natural gas. While Norway is not an EU Member State, its membership in the European Economic Area means it is part of relevant EU regulatory frameworks (Goldthau, 2013). At the same time, Russia is not part of such regulatory bindings and thus demands special attention.

Dependency on Russian natural gas imports varies by EU Member State. Some countries such as the United Kingdom and Romania have low imports while others such as Finland, Latvia, Lithuania, Estonia, Slovakia and Bulgaria are completely dependent on Russian imports (de Vos et al., 2014). In Fig. 3.3, EU Member States are evaluated based on their dependence on Russian gas as well as their ability to adapt to Russian natural gas disruptions. Countries in red either depend on Russian gas for more than 80% of their total annual consumption or are expected to become more dependent. Countries in green have no formal contracts with Gazprom. Those in purple import Russian gas but are protected from disruptions due to sufficient storage capacity, strong relations with Russia, domestic supplies, LNG supplies or because Russian gas plays a marginal role in their energy mix. Gas delivery contracts are typically negotiated bilaterally and not on the European level. Negotiating power therefore varies strongly between the Member States.

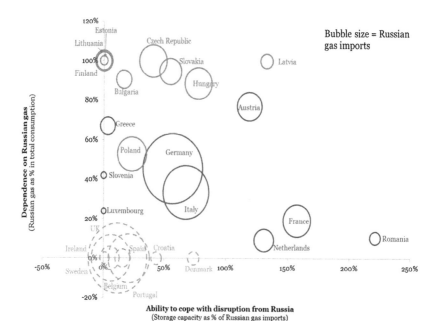

FIGURE 3.3 Energy security of gas imports from Russia in Europe. *Chyong, C.-K., Tcherneva, V. March 17, 2015. Europe's Vulnerability on Russian Gas. Retrieved from:* http://www.ecfr.eu/article/commentary_europes_vulnerability_on_russian_gas.

Note: Due to lack of information, Malta and Cyprus have been excluded from the figures. These two countries consume little gas and receive no gas from Russia.

3.3 EUROPEAN UNION–RUSSIA GAS RELATIONS

Gas trade relations between Russia and the EU can be traced back to the end of the 1960s. The institutional frameworks of both EU and Russian gas markets were defined by national vertically integrated monopolies. This institutional harmony began to diverge in the early 1990s. The fall of the USSR gave rise to the emergence of the vertically integrated joint stock company Gazprom, an export monopoly which then assumed responsibility for EU contracts made with the USSR. The EU began to liberalise the gas industry and develop a single market, most notably through the First and the Second EU Gas Directives (1998/30/EC and 2003/55/EC) and the Third Energy Package, leading to the vertical unbundling of the EU gas sector (Boussena and Locatelli, 2013). Russia began rejecting market liberal energy governance in the 2000s, effectively deviating from the emerging EU model (Kuzemko, 2014). Upon taking on the presidency of Russia in 2000, Vladimir Putin adopted a state-dominated organisation model which contradicted EU reforms to liberalise and unbundle network industries (Boussena and Locatelli, 2013). This has put a strain on relations as the once harmonious gas infrastructures of Russia and the EU have now significantly diverged.

Although attempts have been made to build stronger energy relations between the EU and Russia, the now distinctive aspects of their energy systems complicate such endeavours. In an effort to foster cooperation in the energy sector with former Soviet Union countries and Eastern Europe, the EU signed the European Energy Charter in 1991 ('European Energy Charter', 2007). This was also an attempt to prompt Russia to adopt market principles in the energy sector while assuring supply security and demand reliability in the region (Bilgin, 2011). This multilateral investment treaty would provide guaranteed investments in upstream oil and gas sectors for international oil companies through the establishment of trade, transit and investment rules to liberalise energy flows and investments. Russia signed the Energy Charter Treaty in 1994, initially believing that liberalisation and transition to a market economy could counteract the negative consequences of the dissolution of the USSR, but did not actually ratify it (Bilgin, 2011). There are several reasons for this. Generally, this highlighted Russian resistance to energy sector market liberalisation. On economic grounds, Russia argued against the double standard created by blocking Gazprom from purchasing EU energy companies and the idea that the treaty's transit protocol would not apply between European countries, the EU defining itself as one economic entity (Bilgin, 2011). More detailed arguments against Russia's ratification are seen in the establishment of its non-discrimination principle. The current Russian gas infrastructure is built on dual prices by which gas is sold at a lower price to the domestic market than to the export market. Under the discrimination principle of the Energy Charter, this differentiation in pricing could be viewed as a way to provide hidden subsidies (Boussena and Locatelli, 2013). To better understand the Russian gas industry and the rationale behind its differences of opinion with the EU, it is important to recognise not only the role of natural gas in the Russian economy but also the extent to which it is tied to Gazprom.

3.3.1 Natural Gas and Its Economic Role in the Russian Economy

Following the institutional changes in the Russian gas market, Gazprom emerged as Russia's main gas supplier. As a vertically integrated company, it is involved in exploration and development of gas fields, gas distribution and storage, as well as sales to the final customer. Russia had the second largest proven reserves of natural gas worldwide at the end of 2014 with 32.6 trillion m^3, which equalled nearly 18% of proved global natural gas reserves. Only Iran commanded slightly larger reserves (34 trillion m^3) (BP, 2015a).

Fig. 3.4 shows the development of natural gas production and consumption in Russia since 1990. In 2014 Russia produced 578.7 bcm of natural gas. In the same year gas consumption was at 409.2 bcm (BP, 2015a).

In 2014, Russia exported 187.4 bcm of natural gas via pipeline and further 14.5 bcm as LNG. According to Gazprom, its share in the global and Russian gas reserves is 17% and 72% respectively (Gazprom, 2015a). Additionally, nearly all of Russian gas exports go through Gazprom (Bochkarev, 2013). The Russian state holds slightly more than 50% of Gazprom's stocks (Gazprom, 2015c).

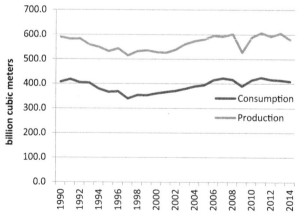

FIGURE 3.4 Natural gas consumption and production of the Russian Federation from 1990 until 2014. *Data from: BP, 2015a. BP Statistical Review of World Energy June 2015. Retrieved from:* http://www.bp.com/content/dam/bp/pdf/Energy-economics/statistical-review-2015/bp-statistical-review-of-world-energy-2015-full-report.pdf.

The role of natural gas for the Russian state is twofold. Gas revenue streams and the connected tax income are an immensely important economic factor. Gazprom contributes 8% to Russian economic output (EurActiv, 2015), pays a quarter of all income tax, and supplies half the country with energy (Müller-Kraenner, 2008). Profiting from maximising gas sale revenues, Russia facilitates Gazprom's corporate expansion (Bilgin, 2011) by trying to reduce Gazprom's exposure to transit countries and by supporting infrastructure ownership, as well as control of and expansion into high profit markets (Orttung and Overland, 2011).

The main profit source for Gazprom is the European market, making Russia dependent on reliable EU gas demand. In 2014, 64% of Russia's natural gas exports went to the EU (BP, 2015a), which equalled 30% of natural gas sales volumes in bcm of Gazprom (Gazprom, 2014a). However, the share of turnover sourced in Europe is much higher than the share of gas volume sold (Boussena and Locatelli, 2013). Due to a dual price system applied for natural gas, differentiating between low domestic gas prices and high export prices, Gazprom makes its profits abroad while posting losses at home (Müller-Kraenner, 2008). The average natural gas price charged by Gazprom within Russia was USD 96.8 per 1000 m3 in 2014. In the same year, the price charged from former Soviet Union countries was at USD 228.5 per $1000\,m^3$. The average price for other exports, summed up in the category 'far abroad', mainly including EU countries, was at USD 297.6 per $1000\,m^3$ (Gazprom, 2014a). Following a simplified estimation on the basis of Gazprom's sales volume in bcm and price data, gas sales to the EU made up approximately 48% of Gazprom's overall sales income of EUR 59 billion in 2014 (net of VAT, excise tax and custom duties) (Gazprom, 2014a).

3.3.2 Natural Gas as a Political Tool

Due to the intimate relationship between Gazprom and the Russian state, the country can leverage its energy supply through Gazprom to reach its political goals. The gas-related foreign energy policy toolbox of Russia contains instruments such as subsidies and differential pricing in favour of strong partners, purchase of infrastructure and assets, pipeline shutoffs, pipeline explosions and construction of alternative pipelines. In addition, there are nonenergy tools, such as military strategies, the creation of custom unions, import bans on nonenergy goods, court cases, rhetoric and public relations campaigns (Orttung and Overland, 2011). Allies can be rewarded with minority shares in companies, pipelines and gas fields (Kropatcheva, 2014).

On the other hand, use of energy power as a foreign energy policy tool comes at a cost, including image damage, mistrust and diversification efforts on the side of the customers (Kropatcheva, 2014). This is particularly relevant to the EU–Russian relationship. In light of recent political realities, most notably Gazprom shutting off gas to Ukraine in January 2009 (Pirani et al., 2009), the

EU has become especially driven to diversify its energy supply and increase its security of supply. However, although the future natural gas demand in the EU is unclear, it is likely that Russia will continue to play a prominent role as a natural gas supplier to the EU. Therefore, it will be essential to establish a fair and functioning natural gas market between the EU and Russia.

3.4 EU POLICIES TARGETED TO IMPROVE NATURAL GAS SECURITY

The EU is focused on preventing threats to its future energy security in the natural gas sector. It is therefore pushing policies to improve its security of natural gas supply through diversification of gas sources and changing market and pricing rules as well as through the increase of renewable energy supply and energy efficiency. In some ways, these policy interventions contradict and challenge the idea of liberal energy markets in Europe, but not in a coherent direction. The irony of these EU market interventions underpins conflicts with Russia, given EU attempts to encourage energy market liberalisation (Kuzemko, 2014).

3.4.1 Energy Union

Due to the economic importance of gas sales to the EU, Russia is dependent on steady and reliable gas demand from Europe, which theoretically gives the EU a strong negotiating position. However, gas delivery contracts are typically negotiated bilaterally, not on the European level and negotiating power varies strongly between the Member States. The result are different gas prices paid by the Member States (Table 3.3). The EU has made efforts to develop comprehensive energy legislation. However, the process of establishing an EU energy structure is slow with many Member States with diverse interests and objections to propositions. In 2015, the European Commission published a communication calling for a European Energy Union in order to strengthen EU energy security. Although the EU has set energy rules at the European level, in practice energy is regulated by 28 national regulatory frameworks, creating inconsistencies and inefficiencies. This communication outlined several strategies to reinforce EU energy security through varying approaches (European Commission, 2015a). First, it proposed changes to the current EU energy market and regulatory structure to better insolate the EU from energy security threats.

The EU Commission has proposed that it takes on the role of central commodity manager and negotiates gas contracts on behalf of Member States (Goldthau and Boersma, 2014). In addition, Member States should inform the EU Commission of future intergovernmental agreements (IGAs) which can then make suggestions to assure a consistent voice across the EU. Full enforcement of the third Internal Energy Market Package and the redesigning of the electricity market to be better integrated throughout the EU are also mentioned as imperative (European Commission, 2015a).

TABLE 3.3 Price Estimations for Russian Gas Bought by EU Countries

	2013	2012	2011
		$ Per 1000 Cubic Metres	
Austria	402	394	387
Bulgaria	394	435	356
Czech Republic	400	500	419
Denmark	382	394	480
Finland	367	373	358
France	404	398	399
Germany	366	353	379
Greece	469	475	414
Hungary	418	416	383
Italy	399	438	410
Netherlands	400	346	366
Slovenia	396	400	377
Slovakia	438	428	333
Switzerland	378	333	400
Poland	429	433	420
Romania	387	424	390

Interfax, November 3, 2014b. Gazprom Paves Way for South Stream With Discounts.

However, the challenge of coordinating very diverse Member State policies and preferences is immense. Furthermore, some observers argue that decentralised contract negotiations between Member States and Russia could actually deliver more optimal pricing than a European gas purchase vehicle. Moreover, a central negotiator could potentially be contradicting the liberalisation of European gas markets that has taken place in the past decades (Goldthau and Boersma, 2014). It might also not be compatible with European competition law (Götz, 2014). The struggle to create a unified energy market will continue to present challenges to the development of an enforceable and unified EU energy security strategy (European Commission, 2015a).

3.4.2 Natural Gas Pricing

Gas delivery contracts between Russia and European Member States have historically been concluded in the form of long-term take-or-pay (TOP) contracts.

TOP contracts have the advantage that price and volume risk are shared between producer and consumer. This is a result of features such as price indexation clauses (mainly to oil products), flexibility clauses and minimum take-off volumes (Boussena and Locatelli, 2013). Usually TOP contracts include a payment guarantee, which obliges the buyer to pay the agreed volume even if no product is taken. In some cases, the buyer has the right to take the gas paid for but not received in subsequent years, either free or for an amount to reflect changes in indexed prices. The long-term nature of the gas delivery contracts enabled producers to invest securely into the development of resources and export infrastructure. TOP contracts harmonised with the former system of vertically integrated energy monopolists in Europe and the Soviet Union (Boussena and Locatelli, 2013).

In the frame of the EU energy market liberalisation process, TOP contracts are being reexamined and being challenged by the EU. According to the EU, TOP contracts create market entry barriers and oppose free-market principles. Currently, old long-term TOP contracts and a spot market where natural gas and LNG prices are determined by supply and demand exist in parallel within the EU. Recently, there have been considerable debates between EU customers and Russia about how to include the price changes on the spot market into the price formula of the TOP contracts (Boussena and Locatelli, 2013). Gazprom is reluctant to changes in the price formula, since the inclusion of spot market prices would have considerable effects on the price and volume risk of Russian gas exports.

Beyond such Gazprom opposition, making changes to current TOP contracts which are legally binding and subject to international arbitration is very difficult. Although the level of TOP for many contracts between Russia and the EU was reduced from 85% to 70% after 2008, further volume reductions or the early termination of contracts is severely limited, with contract durations ranging from 10 to 35 years. Despite European interest to seek other sources of natural gas, Europe is contractually obliged to pay for a minimum volume of Russian gas per year while Gazprom is obliged to provide the promised natural gas volumes to Europe. Consideration of current TOP contracts is therefore critical when examining the relationship between the EU and Russia in terms of natural gas. Assuming a TOP rate in European supply contracts of 70%, European buyers have committed to purchasing 125 bcm of natural gas from Gazprom in 2020 and approximately 70 bcm in 2030.

Fig. 3.5 reflects the original TOP rate of 85% as well as the reduction to 70% in many contracts which came into effect after 2008. It will be particularly important to see if countries with contracts set to expire in the coming years will extend their contracts with Russia, and if so in which form, or whether they will seek alternative natural gas supplies (Dickel et al., 2014).

Furthermore, the Energy Charter, which has been signed but not ratified by Russia, would – once in effect – force Russia to abolish its gas pricing model and to align domestic and export prices, as they could otherwise be charged

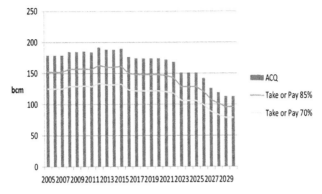

*Data in Russian units; not including Baltic and south East European countries (aside from Turkey and Greece)

FIGURE 3.5   Russian long-term export contracts with OECD European countries to 2030: annual contract quantity and take-or-pay levels. *Henderson, J., Pirani, S., (Eds.) (2014). The Russian Gas Matrix: How markets are driving change, Oxford: OIES/OUP. (as cited in Dickel, R., Hassanzadeh, E., Henderson, J., Honoré, A., El-Katiri, L., Pirani, S., Rogers, H., Stern, J., Yafimava, K., 2014. Reducing European Dependence on Russian Gas: Distinguishing Natural Gas Security From Geopolitics. The Oxford Institute for Energy Studies, Oxford, UK. Retrieved from: http://www.oxfordenergy.org/wpcms/wp-content/uploads/2014/10/NG-92.pdf.)*

of using subsidies to support their domestic market (Boussena and Locatelli, 2013). Gazprom has tried to increase domestic wholesale prices for industrial consumers to bring them closer to the export level and closer to profitability (Gazprom, 2014b). However, the Russian Government usually intervenes, possibly worried about potentially destabilising economic, social and political repercussions and a decrease in industrial output.

The average natural gas price for gas sold by Gazprom 'far abroad', which includes the countries of the EU, was at USD 297.6 per 1000 m^3 in 2014 (USD 8.43 per million Btu) (Gazprom, 2014a). In the same year, domestic customers paid on average around one-third of this price (Gazprom, 2014a), a price that is below cost coverage for Gazprom (The Moscow Times, 2014). Since Gazprom's contracts peg the gas price to the price of oil, the plunge in oil prices between June 2014 and January 2015 could have significant implications on the average gas export price in 2015, which is projected to decrease to USD 222 per 1000 m^3 (USD 6.29 per million Btu) (EurActiv, 2015), thus reducing Gazprom's income streams. Gazprom is therefore finding it more and more difficult to continue supporting the low domestic prices.

3.4.3 Limiting Russian–European Union Market Access

As part of the liberalisation of the gas market in Europe, new organisational models and regulatory standards were introduced by the European Commission that are (partly) opposing the market structures in Russia. The Third Energy Package, adopted in 2009, came into force in 2011 as a means to develop Europe's internal energy market. It includes Directive 2009/73/EC, concerning common rules for the internal gas market. The Directive contains an unbundling provision to separate vertical steps in the gas supply chain, meaning that the company who owns and operates the transmission assets has to be separated from further business activities in the system such as retail or production and import. Vertical integration of transmission system operators (TSOs) was identified as one major obstacle to a well-functioning gas market environment by the European Commission, creating conflicts of interest and market entry barriers. However, several transmission system operators and countries, eg, Germany and France, opposed the unbundling efforts, as they were unsure about the economic benefits and raised juridical arguments against it (Growitsch and Stronzik, 2014).

Following the Directive, the separation can be implemented by the Member States in three ways: ownership unbundling, the introduction of an independent system operator (ISO) or legal unbundling. Under ownership unbundling, all integrated energy companies sell off their gas networks. In this case, no supply or production company is allowed to hold a majority share or interfere in the work of a transmission system operator. With the second option, energy supply companies may still formally own gas transmission networks but must leave the entire operation, maintenance and investment in the grid to an independent company. Under the third option energy supply companies may still own and operate gas or electricity networks but must do so through a subsidiary. All important decisions must be taken independent of the parent company. Ownership unbundling is the strictest unbundling provision and the solution most strongly favoured by the European Commission (Growitsch and Stronzik, 2014).

Nevertheless, Gazprom has carried out asset swapping with some EU Member States, which grants the respective Member States a share in upstream sources and Gazprom access to downstream markets in the importing countries. These agreements are somewhat in contradiction with the rules of the Third Energy Package and based on bilateral agreements between Russia and countries such as Germany, France, United Kingdom and Italy (Boussena and Locatelli, 2013).

Due to the 'third country clause' of the Energy Package, companies from third countries will have to adhere to the same requirements as EU companies. It gives national regulators within the EU the right to deny certification to a TSO that is controlled by an individual or group from a non–EU Member States unless particular requirements are met (Cottier et al., 2010). This clause, in combination with EU ownership unbundling, prevents Gazprom from utilising a downstream integration strategy. However, vertical integration has been one strategy of Gazprom to deal with increased competition on the EU market and to secure its market share. On these grounds, Russia filed a lawsuit with the World Trade Organisation (WTO) in 2014, challenging the EU's Third Energy

Package on the grounds that it goes against WTO basic principles of market access and nondiscrimination (Interfax, 2014a). So far no decision has been made on the issue (World Trade Organisation, 2015).

3.4.4 Diversification of Gas Resources

3.4.4.1 Increasing Imports From Norway

Over the past decade, the patterns of Russian and Norwegian gas supplies to the EU have slightly changed. From 2004 until 2012 the amount of imported Russian gas decreased, mainly due to a reduction of imports from Germany, Italy, Hungary, France and Slovakia. In Germany and France, this was also accompanied by higher imports from Norway. Norway has nearly consistently raised natural gas production since the mid 1990s. The increase in European imports from Norway between 2004 and 2012 stems to a large extent from an increase in imports from the United Kingdom to compensate for their domestic production decline. To a lesser extent, the Netherlands also contributed with an increase in Norwegian imports.

Thus, the overall data suggests a slight shift in EU gas imports towards Norway. However, in 2014, supplies shifted once more away from Norway and towards Russia (Table 3.2, Figs 3.6 and 3.7).

3.4.4.2 Liquefied Natural Gas

Increasing the share of LNG in the EU energy supply has been proposed as an additional strategy towards improving EU energy security by giving the EU leverage for its plans of market liberalisation and changes to contract conditions with parties such as Russia. However, despite some seeing LNG imports as playing a large role in diversifying EU natural gas supplies, in practical terms, LNG imports have yet to prove themselves to be significant in real markets.

As most LNG is sold under long-term contracts which are typically linked to the oil price, there is a direct connection to the recent plunge in oil prices causing substantial price decreases for LNG. As Table 3.5 shows, the average price of crude oil was at USD 16.80 per million Btu in 2014, hovering below USD 10 per million Btu in July 2015. At the same time the Japan LNG import price fell from USD 16.13 per million Btu end of June 2014 to 8.50 end of June 2015 (YCharts, 2015a). Without signs of notable price recovery, oil and gas companies are likely to slash capital expenditure programmes and shrink their 2015 budgets. Consequently, they will focus on core assets instead of investments in new ventures. Expensive, less-profitable projects are especially vulnerable and many are set to be cancelled in response to market conditions (International Energy Agency, 2015). Those projects which have already invested the substantial upfront costs for LNG plants will come online as planned but current LNG prices lack the capital return to merit the high capital costs of building new plants. Should prices stay low, 2020 may see a significant shrinking of LNG

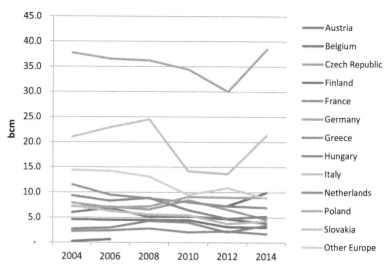

FIGURE 3.6 Recent development of natural gas pipeline imports from Russia. *Data from: BP, 2005. BP Statistical Review of World Energy June 2005. BP. Retrieved from:* http://www.bp.com/ statisticalreview; *BP, 2007. BP Statistical Review of World Energy June 2007. BP. Retrieved from:* http://www.bp.com/statisticalreview; *BP, 2009. BP Statistical Review of World Energy June 2009. BP. Retrieved from:* http://www.bp.com/statisticalreview; *BP, 2011. BP Statistical Review of World Energy June 2011. BP. Retrieved from:* http://www.bp.com/statisticalreview; *BP, 2013. BP Statistical Review of World Energy June 2013. BP. Retrieved from:* http://www.bp.com/statistical-review; *BP, 2015a. BP Statistical Review of World Energy June 2015. Retrieved from:* http://www. bp.com/content/dam/bp/pdf/Energy-economics/statistical-review-2015/bp-statistical-review-of-world-energy-2015-full-report.pdf.

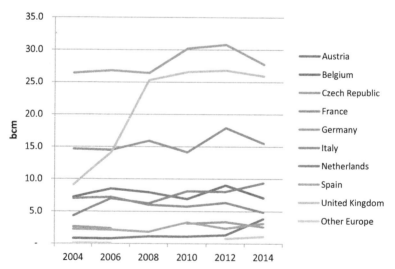

FIGURE 3.7 Recent development of natural gas pipeline imports from Norway. *Data from: BP, 2005. BP Statistical Review of World Energy June 2005. BP. Retrieved from:* http://www.bp.com/statistical-review; *BP, 2007. BP Statistical Review of World Energy June 2007. BP. Retrieved from:* http://www. bp.com/statisticalreview; *BP, 2009. BP Statistical Review of World Energy June 2009. BP. Retrieved from:* http://www.bp.com/statisticalreview; *BP, 2011. BP Statistical Review of World Energy June 2011. BP. Retrieved from:* http://www.bp.com/statisticalreview; *BP, 2013. BP Statistical Review of World Energy June 2013. BP. Retrieved from:* http://www.bp.com/statisticalreview; *BP, 2015a. BP Statistical Review of World Energy June 2015. Retrieved from:* http://www.bp.com/content/dam/bp/pdf/Energy-economics/statistical-review-2015/bp-statistical-review-of-world-energy-2015-full-report.pdf.

markets in contrast to growth expectations. The final investment decisions made in the next two years will decide available incremental LNG supplies at the start of 2020 (International Energy Agency, 2015).

Australia, seen as a major LNG export up-and-comer along with the Untied States, has suffered extensive cost overruns in the past and is reaching an overall LNG cost of USD 14–16 per million Btu, making their projects less profitable. Even if investors will treat their investment as sunk costs and sell gas without profits, the EU would most likely not profit from these exports as destination clauses in the Australian contracts leave Australian LNG exports bound to the Asian market alone (Natural Gas Europe, 2014). The importing of LNG from the United States is seen as a possible means to diversify European energy supply. In fact, because US LNG plants do not include destination clauses and US gas prices are delinked from oil prices, US LNG is particularly desirable as well as flexible. Theoretically, the United States could deliver gas to the EU at USD 10 per million Btu (Natural Gas Europe, 2014).

On the other hand, low LNG prices lead to an increase in demand and open up new markets. Furthermore, they could incentivise innovations and cost reduction efforts, such as miniaturisation (reduced project scale and complexity) or modularisation (distribution of fabrication to low-cost regions and on-site assembly). Thus long-term cost reductions could be achieved (Stoppard, 2015).

In the EU, LNG imports have been declining dramatically over the past years. From 2011 to 2014, European LNG imports have decreased by 47% (BP, 2013, 2014, 2015). Table 3.4 shows the amount of LNG imports from 2010 to 2014.

There are several reasons for this decline. The US shale-gas boom, coupled with slowing Chinese demand, put downward pressure on coal prices to make them cheaper than natural gas ('The unwelcome renaissance', 2013). From 2008 to 2013, the average price of coal in north-west Europe fell by 45%. This resulted in a 9% spike in EU coal demand from 2011 to 2012 compared to 2010 levels (The Economist Intelligence Unit, 2014). Although EU coal consumption dropped again by approximately 9% in 2014 from 2012 levels, this, along with

TABLE 3.4 European Liquefied Natural Gas Imports in bcm From 2010 to 2014

Year	2010	2011	2012	2013	2014
LNG imports in bcm	80	85	62	45	45

BP, 2011. BP Statistical Review of World Energy June 2011. BP. Retrieved from: http://www.bp.com/statisticalreview; BP, 2012. BP Statistical Review of World Energy June 2012. BP. Retrieved from:http://www.bp.com/statisticalreview; BP, 2013. BP Statistical Review of World Energy June 2013. BP. Retrieved from: http://www.bp.com/statisticalreview; BP, 2014. BP Statistical Review of World Energy June 2014. BP. Retrieved from: http://www.bp.com/statisticalreview; BP, 2015a. BP Statistical Review of World Energy June 2015. Retrieved from: http://www.bp.com/content/dam/bp/pdf/Energy-economics/statistical-review-2015/bp-statistical-review-of-world-energy-2015-full-report.pdf.

greater renewable energy generation capacity and low carbon prices, had implications for economic investment in gas-fired power plants (BP Review, 2015; International Gas Union, 2014).

Furthermore, the price spread between LNG and EU pipeline gas (approximated by average German natural gas import price) increased significantly, as can be seen in Table 3.5, giving less incentive to purchase LNG on the market. Only very recently has this trend turned.

Despite the optimism surrounding LNG potential, there are still many challenges to meaningful LNG development in addition to the current less-than-ideal market conditions. Companies would first have to go through complicated and costly licensing procedures to establish export terminals. Even if this process was circumvented, billions of dollars would then be needed to either build new greenfield LNG export facilities or convert existing terminals (Goldthau and Boersma, 2014).

Though the EU market may be able to receive LNG supplies, these come at a substantial cost (Goldthau and Boersma, 2014). The cost of shipping US LNG to Europe is very high with the US finding it difficult to compete with more nearby LNG producers such as Qatar, Norway and Algeria in the current market. Conversely, it is possible that as the US supplies other markets, more LNG from closer sources such as North Africa, Qatar or the Persian Gulf might become available (Kropatcheva, 2014), which could also lead to reduced prices.

TABLE 3.5 Development of Natural Gas and Liquefied Natural Gas Price Indicators from 2010 to 2014

USD Per Million Btu	2010	2011	2012	2013	2014	July 2015
Crude oil OECD (cif)	13.47	18.56	18.82	18.25	16.80	9.79[a]
Natural gas average German import price (cif)	8.01	10.49	10.93	10.73	9.11	6.93[b]
Liquefied natural gas (LNG) Japan import price (cif)	10.91	14.73	16.75	16.17	16.33	8.50[c]
Spread LNG pipeline gas	2.90	4.24	5.82	5.44	7.22	1.57

[a]average crude oil spot price (YCharts, 2015b).
[b]YCharts (2015c).
[c]YCharts (2015a).
Adapted from: BP, 2015a. BP Statistical Review of World Energy June 2015. Retrieved from: http://www.bp.com/content/dam/bp/pdf/Energy-economics/statistical-review-2015/bp-statistical-review-of-world-energy-2015-full-report.pdf.

As far as the ability for US LNG to compete with Russian natural gas imports goes, the US is at a large disadvantage. Transportation via pipeline is much cheaper and the infrastructure is already there. If anything, US LNG can provide the EU with leverage when it tries to modify its contract terms with Russia (Rogozhin, 2015).

Additionally, some in Europe are hopeful that US shale gas can be directly imported in the short-term. To do this, substantial limitations are in place set out by American laws and regulations which restrict LNG export to countries without a free trade agreement and require long-term commitments to license an LNG export terminal. Infrastructure is still in its infancy with six approved export terminals all still under construction as of January 2016 (FERC, 2016).

The European gas market is also poorly integrated, making the easy flow of natural gas through scattered national markets difficult. Infrastructural limitations and pipeline network design restricts the total volume of LNG transit that can take place within Europe. The lack of market harmonisation and variations in prices due to direct state influence outside of north-west Europe add to a system that is unattractive to foreign investors (Goldthau and Boersma, 2014). Further development of small-scale LNG (SSLNG) also demands the creation of a consistent normative and regulatory framework which also outlines safety standards for SSLNG handling (International Gas Union, 2014). The EU Commission has stated it will prepare an LNG strategy to improve needed infrastructure and remove LNG import obstacles between it and LNG producers such as the United States but its successful development and implementation has yet to be done (European Commission, 2015a).

3.4.4.3 Pipelines and Interconnectors

To further diversify its natural gas supply, the EU has proposed the construction of various pipelines which would allow natural gas to flow to the EU without crossing Russian territory. A major EU initiative to this end is the construction of the Southern Gas Corridor (European Commission, 2014a) primarily consisting of the South Caucasus Pipeline (SCP), Trans-Adriatic Natural Gas Pipeline (TAP) and the Trans-Anatolian Natural Gas Pipeline (TANAP) project (Trans Adriatic Pipeline, 2015; see Fig. 3.8).

The Southern Gas Corridor marks itself as one of the most elaborate gas value chains ever attempted. Its proposed 3500 km will wind through seven countries and consist of numerous separate energy projects, all to the tune of about USD 45 billion in total investment. The three pipeline projects are of notable interest. The SCP crosses Azerbaijan and Georgia while the South Caucasus Pipeline Expansion (SCPX) is underway to, among other things, connect SCP to TANAP. TANAP will be focused on Turkey. Finally, the TAP will involve the countries of Greece, Italy and Albania. In particular, TAP brings diversity to natural gas transportation by providing additional opportunities to connect with current and proposed pipelines to better connect Europe.

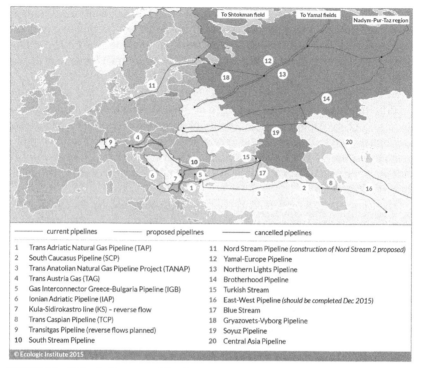

FIGURE 3.8 Main pipeline routes delivering gas to the European Union. *Ecologic Institute 2015: own depiction.*

Notably, its connection to Italy can then provide gas to the rest of Europe. It also expands the ability for Europe to use reverse flows. Gas from TAP can be carried to Eastern and Central Europe through the Trans-Austria Gas (TAG) pipeline by utilising swaps and reverse flows. Building the capacity to use reverse flows through the current Transitgas pipeline via Switzerland could add another possibility to bring gas to Germany and France. Further reverse flows have been planned to go between Italy and the UK. TAP offers new gas sources to Bulgaria in addition to the capability to send reverse flows to the Kula-Sidirokastro line and the planned Gas Interconnector Greece–Bulgaria (IGB) pipeline. In collaboration with the planned Ionian Adriatic Pipeline (IAP), opportunities are also being discussed for TAP to help bring supplies to markets like Albania, Montenegro and Bosnia and Herzegovina which do not have gas (Trans Adriatic Pipeline, 2015).

The goal of the Southern Gas Corridor would be to eventually transport gas from the Shah Deniz II field in the Caspian Sea through Turkey before reaching the EU. To begin, initial capacity would support 10 bcm per year with full capacity being reached in 2026 at 31 bcm per annum and 60 bcm by 2030 (Natural Gas Europe, 2015b). A key advantage of the pipeline is the addition of underground storage facilities and a national gas grid in Albania to augment the reliability of gas supplies in times of disruption. The pipeline is argued to be cost-effective for its potential to connect with other planned pipelines, such as the IAP or IGB (Natural Gas Europe, 2015c).

Another proposed pipeline is the Trans-Caspian Pipeline (TCP) to link Turkmenistan and Azerbaijan by the Caspian Sea. By the end of 2015, the East–West Pipeline is projected to be completed by Turkmenistan to connect the Caspian Sea coast to Turkmen gas deposits in the southeast to supply the TCP or the western corridor more broadly with 40 bcm per year (Natural Gas Europe, 2015b). Once the TCP is completed, this gas can then travel through this pipeline and on to reach the EU market (Cutler, 2015).

In addition, the EU is improving its internal market and gas interconnectors. The EU has recognised the urgent need to update and expand its system of interconnections among EU Member States to achieve its energy security goals for its internal energy market. An interconnected energy system offers increased security of supply, more affordable prices in the internal market and is essential for sustainable development and decarbonising its energy mix. Of particular focus has been the deeper integration of Central Eastern and South Eastern European countries into the EU gas market. On July 10, 2015, a memorandum of understanding was signed, formally starting an initiative which will see 15 EU and Energy Community countries in these regions cooperate to build missing gas infrastructure links and solve technical and regulatory issues blocking further development. This stable market and regulatory framework comes in the context of the EU Commission's Central East South Europe Gas Connectivity (CESEC) High Level Group which was established in 2015 to promote regional market integration and natural gas diversification. Along with building new pipelines, strategies for improving the use of current infrastructure, particularly capacity for reverse flows, will be prioritised (European Commission, 2015b). This, along with the development of the 2013 trans-European energy networks regulation (TEN-E) which, for the first time, set out a method for dealing with cross-border projects (EU Commission, 2015), can make strides in future natural gas infrastructure improvements.

3.4.4.4 Gas Storage

Expansion of gas storage capacity has been indicated as a way to bolster EU energy security. In 2010, Regulation No 994/2010 was passed requiring Member States to assure that their gas companies could guarantee gas delivery obligations to customers and survive a complete gas cut-off from their largest single supplier. Under extreme conditions, supplies must be assured for a minimum of 30 days of high demand as well as in the case of a seven day peak in temperature (EUR-Lex, 2011). While this prompted the development of greater storage, the lack of a uniform standard for storage leaves each Member State responsible for determining the level of storage needed. Though the building of an EU strategic gas reserve similar in concept to the US Strategic Petroleum Reserve has been touted as too costly, it is possible to modify gas storage tariff regimes through legislation in order to better incentivise commercial gas storage and increase storage capacity (Buchan, 2014). The Energy Union notes the importance of developing new storage technologies for traditional gas and LNG. To this end, the EU Commission launched a public consultation, running from July to September 2015 to gather information on perceived challenges and opportunities for EU gas and LNG storage as well as how storage might strengthen security of supply (European Commission, 2015d). The results will feed into the development of an EU strategy to realise the full potential of such storage (European Commission, 2015c). As of September 2015, the declared total maximum technical gas storage of the EU amounted to 92.14 bcm (Gas Infrastructure Europe, 2015), which roughly equals a quarter of the yearly natural gas consumption in the EU in 2014.

3.4.4.5 Shift Towards Low-Carbon Policies and Technologies

The introduction of climate targets and policies does not only make sense in terms of climate change mitigation but also in the achievement of energy security objectives (Kuzemko, 2014), since the increase of renewable energy capacity and improvements in energy efficiency can reduce import dependencies. Until 2030 the EU has agreed on a set of targets including 27% of renewable energy in final energy consumption (binding at EU-level) and an indicative target of 27% energy efficiency improvements compared to projections of future energy consumption (European Commission, 2014b). Until 2020 the target framework from the Climate and Energy Package will remain in place. While it might be debatable whether the targets implemented are stringent enough or their legal status appropriate, their fulfilment will have implications on European energy security. In accordance with the target frameworks, there is a variety of climate policies in place to reduce greenhouse gas (GHG) emissions as well as support and enforce not only renewable energy deployment but also energy efficiency measures.

Fig. 3.9 gives an overview about the implemented policies. As far as gas is concerned, the most important sectors to address are the residential, power and industry sectors (compare to Fig. 3.2).

The connection between renewable energy and energy efficiency policies on the one hand and reduced fossil energy consumption and thus lower fossil energy imports on the other hand is very straight forward. However, energy security is also very much affected by the kind of fossil fuel energy carrier used in the energy system.

Coal is an energy carrier that has been more economical than gas over the past decades. Hard coal is available in abundance on the world market and can be transported at relatively low cost (World Coal Association, 2012). Lignite has to be purchased from sources close to the plant. Its low energy content per unit of weight makes long distance transport uneconomic. However, there are

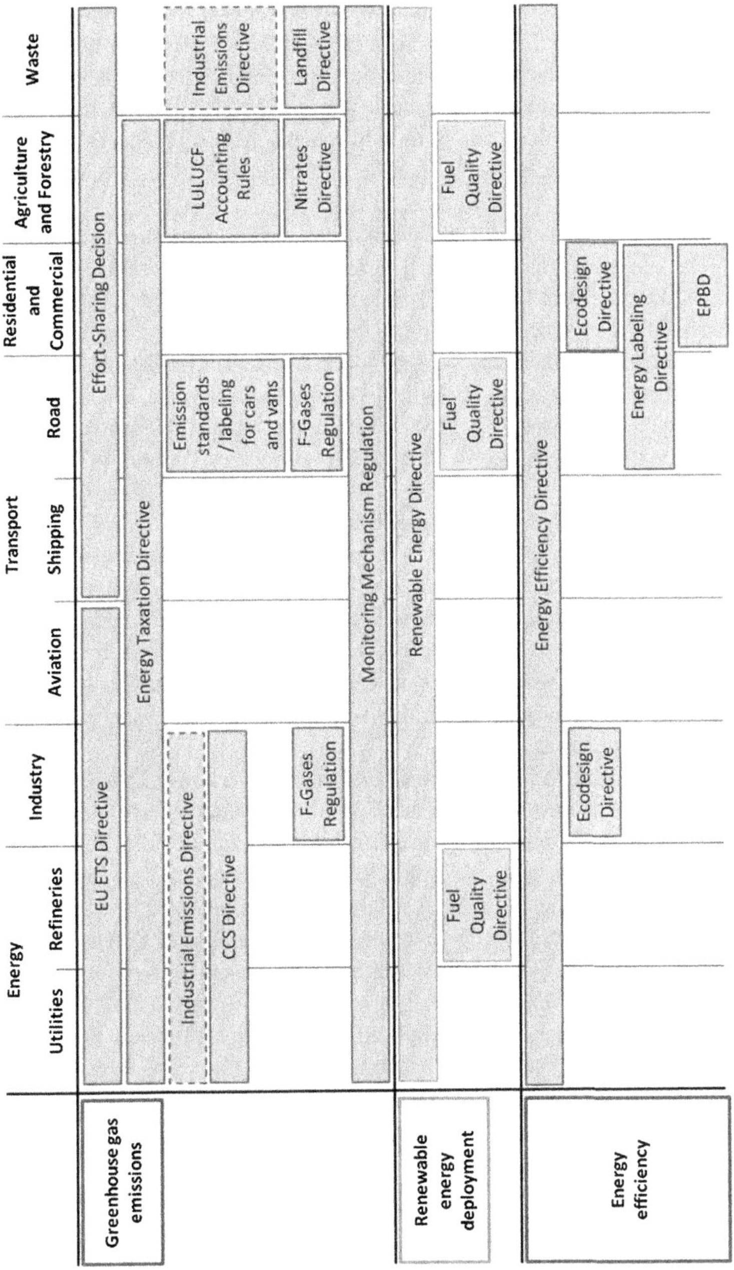

FIGURE 3.9 Climate policies in the EU and sectors affected. *Ecologic Institute 2014: own depiction.*

still significant sources of lignite in the EU. Thus, an increased use of coal could potentially improve energy security in combination with renewable energy and energy efficiency measures more than would be seen in a renewable energy–natural gas combination. However, an increased use of coal as a high carbon energy carrier would completely undermine the GHG reduction targets of the EU. A fulfilment of climate and energy security objectives is therefore most likely to be achieved with a combination of renewable energy, energy efficiency and gas serving as a 'bridge technology' as part of the transition to a low-carbon economy (Lazarus et al., 2015).

If we take a look at the role of natural gas in the future energy system of the EU, the European Commission gas demand projections are an interesting source to examine. Since 2003 these projections had to be reduced for each new reference scenario produced since 2003. Over the past 10 years, projections for 2015 have dropped by 23% and yet the Commission still predicts that 2015 gas demand will increase by 20% from 2014 levels (Jones et al., 2015). In 2014, demand in the EU dropped by nearly 12% compared to 2013, returning to levels not seen since the mid 1990s (BP, 2015a). The European Network of Transmission System Operators for Gas (ENTSO-G), responsible for planning the EU gas pipeline network, has also proposed what turned out to be inflated natural gas demand numbers. Contrary to its projected 8% increase in natural gas demand from 2010 to 2013, demand actually fell by 14% (Jones et al., 2015).

Fig. 3.10 shows the development of long-term gas demand forecasts, which have been revised downwards since 2003. Compared to current trends, warnings have been given that overestimations of natural gas demand for the EU could lead it in the wrong direction, investing money in excessive natural gas infrastructure (Jones et al., 2015).

However, there are various other energy system projections in place. Table 3.6 shows projections from the European Commission's Energy Roadmap 2050, displaying the modelling results for the Current Policy Initiatives (CPI) scenario, the High Renewable Energy source (RES) scenario and the High Energy Efficiency (EE) scenario. The CPI scenario includes policies adopted by March 2010, including the 2020 targets for RES share and GHG reductions as well as the Emissions Trading System (ETS) Directive. Furthermore, it includes measures adopted afterwards, eg after the Fukushima events, and being proposed in the Energy 2020 strategy; it also includes measures proposed in the Energy Efficiency Plan and the new Energy Taxation Directive (EU Commission, 2011a). The High RES scenario includes strong support measures for RES leading to a very high share of RES in gross final energy consumption (75% in 2050). The High EE scenario commits to very high energy savings, realised through measures such as more stringent minimum requirements for appliances and new buildings, high renovation rates of the existing building stock or the establishment of energy savings obligations on all energy utilities. It assumes a significant increase of renewable energy capacity as well. This is expected to be a decrease in energy demand of 41% by 2050 as compared to the peaks in 2005–2006 (EU Commission, 2011a).

European Commission EU27 Gas demand forecast

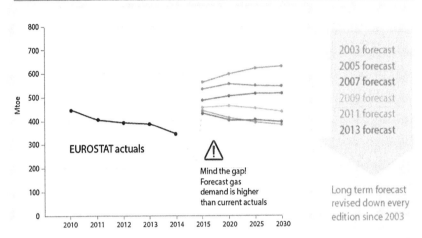

FIGURE 3.10 European Commission EU-27 gas demand forecasts. *Jones, D., Gaventa, J., Dufour, M., 2015. Europe's Declining Gas Demand: Implications for Infrastructure Investment and Energy Security. E3G. Retrieved from:*http://e3g.org/docs/E3G_Europes_Declining_Gas_Demand_10_6_2015.pdf.

Compared to 387 bcm of actual natural gas consumption in 2014 (see Section 3.2), natural gas consumption in the CPI scenario is projected to be at 443 bcm in 2030, to be reduced slightly until 2050. In both the renewable energy and the energy efficiency scenario, the role of natural gas in the energy sector is substantially reduced until 2050, while the share of imports remains equal over all scenarios at around 80% of gas consumption in 2030 and 90% in 2050. The high RES scenario reaches a natural gas consumption reduction of some 4% in 2030 and 41% in 2050 in comparison to the CPI scenario. The energy efficiency scenario comes to a reduction of natural gas consumption until 2030 of 6% and 28% in 2050. Furthermore, both of the more ambitious scenarios reach a reduction of energy use in all other energy carriers, especially in coal in comparison to the CPI scenario.

In a different study conducted by Ecofys in 2014, three scenarios are examined for natural gas demand in the EU until 2030 (van Breevoort et al., 2014). The business-as-usual (BAU) scenario is based on the 2013 PRIMES baseline scenario and includes current economic and demographic trends and existing policies in that year. The 2030 framework for climate and energy policies scenario builds on the initiatives outlined in the EU 2030 framework, such as the renewable energy and energy efficiency target. The energy independence scenario (energy efficiency and renewable energy) includes an increased implementation rate of renewable energy, equal to the growth rates in from 2000 to 2010, as well as further energy efficiency progress (van Breevoort et al., 2014). Fig. 3.11 shows the scenario results.

TABLE 3.6 The Role of Natural Gas and Renewable Energy According to the Energy Roadmap 2050

Energy Roadmap 2050 EU-27	Gross Inland Consumption	2030	2050
Current policy Initiatives scenario	Natural gas (ktoe)	3,69,465	3,54,256
	Natural gas (bcm)	443	425
	Of which net imports	79%	91%
	Renewable energy source (RES) (ktoe)	3,13,796	3,75,464
	RES in final energy consumption	19%	23%
	RES in gross final energy demand	25%	29%
	Solids (ktoe)	1,94,962	1,52,130
	Oil (ktoe)	5,54,930	5,16,760
	Nuclear (ktoe)	1,97,877	2,18,790
High RES scenario	Natural gas (ktoe)	3,53,672	2,10,434
	Natural gas (bcm)	424	253
	Of which net imports	80%	91%
	RES (ktoe)	3,87,095	6,76,202
	RES in gross inland consumption	26%	60%
	RES in gross final energy demand	31%	75%
	Solids (ktoe)	1,08,644	23,406
	Oil (ktoe)	5,14,095	1,75,825
	Nuclear (ktoe)	1,45,795	43,640
High energy efficiency scenario	Natural gas (ktoe)	3,47,506	2,56,714
	Natural gas (bcm)	417	308
	Of which net imports	79%	91%
	RES (ktoe)	3,17,439	4,72,139
	RES in gross inland consumption	22%	44%
	RES in gross final energy demand	28%	57%
	Solids (ktoe)	1,31,659	44,367
	Oil (ktoe)	4,95,657	1,67,513
	Nuclear (ktoe)	1,61,520	1,46,133

Adapted from: EU Commission, 2011b. Energy Roadmap 2050 Impact Assessment Part 2 Including Part II of Annex 1 "Scenarios – Assumptions and Results" and Annex 2 "Report on Stakeholders Scenarios" (Commission Staff Working Paper No. SEC(2011) 1565). EU Commission, Brussels, Belgium. Retrieved from: https://ec.europa.eu/energy/sites/ener/files/documents/sec_2011_1565_part2.pdf.

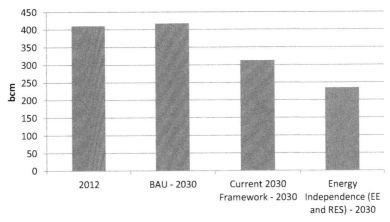

FIGURE 3.11 European natural gas consumption according to different scenarios. *Adapted from: van Breevoort, P., Hagemann, M., Höhne, N., Day, T., de Vos, R., 2014. Increasing the EU's Energy Independence Technical Report (No. CLIDE15208). Ecofys. Retrieved from:*http://www.ecofys. com/files/files/ecofys-2014-increasing-eu-s-energy-independence-technical-report.pdf.

In the BAU scenario, EU natural gas imports increase due to a reduction in solids, oil and nuclear energy in the energy mix while total primary energy consumption remains nearly at 2012 levels. The other two scenarios show significant decreases in gas demand, by 25% and 44% in comparison to BAU-2030 respectively, with nearly the same total energy demand but different levels of renewable energy in the mix (van Breevoort et al., 2014).

In contrast to the scenarios explained before, there are also projections showing increased demand in natural gas in the future. A report by the Natural Gas Programme of the Oxford Institute for Energy Studies indicates that European natural gas demand will increase to 618 bcm in 2030 (Honoré, 2014). The BP Energy Outlook 2035 for the EU states that from 2013 until 2035 oil consumption is projected to fall by 23% and coal by 54%, while natural gas consumption will rise by 15%. In the same time span renewable energy is expected to expand by 136% to reach 18% in primary energy consumption in 2035 (BP, 2015b). In this trajectory the expansion of RES and EE improvements are expected to be lower than in the ambitious scenarios above. The reduction of coal and oil use is therefore compensated with natural gas to reduce overall GHG emissions.

In summary, it can be stated that low-carbon policies and technologies can cause a reduction in natural gas consumption. The role of gas in the future energy system, however, depends on the ambition and stringency of climate targets and policies and their fulfillment.

Nevertheless, it needs to be considered that even under the ambitious RES and EE improvement scenarios, there is still a significant role for gas in a low-carbon energy system. Especially in view of depleting domestic reserves, gas imports are likely to affect EU energy security still at some level in the future.

3.4.4.6 European Union External Policy Instruments

In addition to the Energy Charter Treaty (see Section 3.3) which intended to harmonise the EU and Russian natural gas markets, other EU external policy instruments have been implemented to improve energy security in the EU through a stronger EU–Russia partnership. The basis for EU–Russia collaboration today stems from the 1997 Partnership and Cooperation Agreement to promote EU–Russia political dialogue in addition to economic and financial cooperation (EUR-Lex, 1997). This collaborative relationship has been built upon over the years through other agreements. Established in 2000, the EU–Russia Energy Dialogue has provided a structure for cooperation. The dialogue serves to facilitate information exchanges on legislative initiatives, increased investment in energy efficiency and sustainable energy generation, updated and secure energy infrastructure and greater access and opportunities in energy markets (European Commission, 2015e). Though promoting dialogue between the two parties, it has often been criticised for not translating talk into concrete actions to achieve its goals (Talseth, 2012). One of its greatest achievements, however, was creating the Early Warning Mechanism in 2007 within its framework. The Early Warning Mechanism aims to prevent and quickly respond to an emergency or threat of an emergency regarding the supply and demand of natural gas, electricity and oil through the establishment of practical measures between the two parties. The tools to accomplish these goals include information exchange, notification, consultations, monitoring and early evaluation (Coordinators of the EU–Russia Energy Dialogue, 2011). The 2010 Partnership for Modernisation continues to foster exchange to promote economic and technical modernisation which is directly involved in matters of energy (European Union External Action, 2015). Though the events in Ukraine have strained collaborative efforts, these are just some of the mechanisms which continue to strengthen the EU–Russia relationship as it relates to energy security.

3.5 RUSSIAN GAS STRATEGIES UNDER UNCERTAINTY

The changes in the European market discussed in Section 3.4 have increased the volume and price risks for Russia as a gas supplier (Boussena and Locatelli, 2013). These measures, taken by the EU to improve energy security and to limit Russian influence and market power in the gas market reduce Russia's security of demand. From the Russian perspective, the EU is the main gas customer. Thus, gas import reductions on the side of the EU translate into loss of revenue and political influence for the Russian state. A sharp reduction of its largest export market can clearly have serious economic implications (Kuzemko, 2014).

This insecurity of demand from Europe imposes questions on the side of the Russian government on how to deal with this situation in terms of where to find buyers to compensate for the loss and how much to invest into exploration, production and transit infrastructure. Among efforts to protect its own interests in the face of declining demand security on the EU market, Russia is looking towards the East to gain customers to hedge against risks on the European market.

Implementing its 2030 energy strategy, which envisages a share of 19–20% of pipeline-transported natural gas going to the Asia Pacific markets (Mareš and Laryš, 2012), Russia is developing Eastern resources and gas transmission infrastructure towards the East, such as the Power of Siberia and Power of Siberia 2 pipelines, in the frame of the Eastern Gas Programme (Gazprom, 2015b). In May 2014, Russia and China signed a 30 year gas delivery deal worth USD 400 billion, agreeing on a yearly delivery of 38 bcm of natural gas from Gazprom to China National Petroleum Corporation. This amounts to a price in the range of USD 350 to USD 400 per 1000 m^3 (RT, 2014), which is above the average natural gas price for gas sold by Gazprom to 'far abroad' of USD 297.6 per 1000 m^3 in 2014 (Gazprom, 2014a), as well as the average German import price of USD 322 per 1000 m^3 in 2014 according to BP (BP, 2015a).

Gazprom hopes to ship further 30 bcm of gas a year to China under a nonbinding framework agreement settled in November 2014. However, the decrease in oil prices put pressure on Russian price expectations and lower gas prices might make investments into the necessary costly pipeline infrastructure unprofitable. Moreover, China has some leverage in price negotiations: The competition on the Asian market is higher than in Europe and China is strongly diversifying its energy mix. It has its own pipeline projects (eg, the Central-Asia-China pipeline which will provide gas from Turkmenistan, Uzbekistan and Kazakhstan to China), develops own natural gas resources, expands LNG use, plans to develop domestic shale gas resources and is also heavily enforcing renewable energy (Kropatcheva, 2014; Mareš and Laryš, 2012). There is, therefore, a risk that Russia will not be able to fully substitute revenue from European gas demand with demand from the Asian continent. The turn to the East is also geopolitically less advantageous. The political leverage from gas deliveries will be significantly smaller in China than in the EU, as China's energy supply is highly diversified, making the projects less attractive from a foreign energy policy point of view (Mareš and Laryš, 2012).

Furthermore, Russia has a vested interest in assuring that Russian natural gas production can continue to provide for EU natural gas consumption demand into the future and meet its current TOP contract obligations. However, delays in new field developments and gas investment have caused doubts as to whether Gazprom can actually deliver on these obligations (Nowak et al., 2015). In order to keep Russian gas production at current levels, significant investment is needed and new fields need to be explored, posing serious problems for Gazprom (Kropatcheva, 2014). Production in Western Siberia's Nadym Pur Tazov district is decreasing each year with projections that production will drop from current levels of 500 bcm to 333 bcm in 2035. Yamburg, Urengoy and Medvezhye, three of the district's largest gas fields, are already depleted by over three-quarters with an estimated annual decline of 25–30 bcm to come. New developments in the Yamal region hold the most promise, with production to increase from a few bcm today to over 200 bcm in 2035. Gas production is projected to increase in the Eastern Siberian and Far Eastern regions by 82 bcm and 64 bcm, respectively, by 2035. Further development of the High North's Shtokman field is an important part of maintaining current Russian gas production levels (Nowak et al., 2015).

There are numerous challenges to investing in the further development of Russian natural gas fields which have led to delays in investment projects. Field development requires significant financial investments, which the IEA and Russian Ministry of Energy estimate at annual USD 30 billion and must prove to be economically viable (Nowak et al., 2015). For example, although the Shtokman reserve holds an estimated 4 trillion m^3 of natural gas, it was announced in 2012 that its development would be postponed due to prohibitively high costs which make it unable to receive needed returns on investment for the time being (Bellona, 2012). However, a reduction in security of demand and falling or fluctuating prices increase the risk of new field developments.

Russia must not only assure Russian natural gas production meets EU demand, but it must further guarantee that those supplies reach the EU market. Russian export strategies have focused on diversifying supply routes to bypass transit countries, particularly Ukraine to more easily guarantee uninterrupted gas transport into the EU. This often results in a blending of political and economic interests with Russia having to strike a balance between the two in order to increase EU natural gas import security.

Ukraine has served as a major transit country for EU-bound Russian gas but transit disputes, Ukraine's domestic instability and conflict between the two countries create an unstable relationship. When gas flows through Ukraine were interrupted for two weeks in January 2009, Russia stated losses of USD 2 billion (Hafner, 2012). Although gas flows through Ukraine have been on the decline, Russia continues to be highly dependent on the Ukrainian Gas Transmission System (GTS). As Kiev tries to reduce its Russian gas imports, in part because it considers them too expensive, Russia references minimum gas import volumes agreed to between the two countries. Russia has indicated that cheaper gas to Ukraine would be a possibility if Ukraine sold the Ukrainian GTS to Gazprom (Hafner, 2012).

Belarus is the second most important transit country for Russian natural gas with an estimated 20% of European gas imports from Russia travelling through the country. After the 2004 Russo-Belorussian conflict, Belarus sold control of its GTS, Beltransgaz, to Russia. As a result, prices for domestic gas in Belarus are projected to be comparable to Russian domestic prices. Gazprom is interested in increasing transit through Belarus' Yamal-Europe pipeline as a means to decrease flows through Ukraine and better guarantee supplies to Western Europe. The EU has also indicated that investment in this pipeline should be a top priority (Hafner, 2012).

In an attempt to reduce its dependence on Ukraine as a transit country for its gas, also the Nord Stream pipeline was constructed to connect Russia directly to Germany, its major gas market in Western Europe. The Turkish Stream project could have been another means to circumvent Ukraine as a transit country to bring gas to the EU via Turkey. The project was brought forward to replace the previously planned South Stream project which Gazprom cancelled in 2014 in response to strong political pressure from the EU Commission and particular EU Member States (Natural Gas Europe, 2015a). However, relations between Russia and Turkey have deteriorated on the background of military operations in Syria, which is likely to complicate Turkish Stream implementation (Hafizoglu, 2015).

Furthermore, plans for building stage 3 and 4 of the Nord Stream pipeline, known as Nord Stream II (Zhdannikov and Pinchuk, 2015), reduces the needs for the Turkish route. Nord stream II will parallel the current Nord Stream pipeline and is scheduled to be completed in 2019. Able to transport 55 bcm per annum, the pipeline further helps to divert gas imports to the EU from bypassing Ukraine. It simultaneously boosts Germany's position as the main gas hub in Europe which has been a long-term national priority (Natural Gas Europe, 2015d). Partners include Anglo-Dutch Shell, Austrian OMV and German E.ON. Though relations between the EU and Russia may make approval of the pipeline difficult as it moves towards less dependence on Russian energy, E.ON argues that increased demand for Russian gas in Europe will come in the face of declining domestic production. Besides this, the largest utility in Germany stresses that the construction of this pipeline is the cheapest and most efficient means to ensure European energy security, noting that supply route diversification is just as important as discovering new suppliers (Zhdannikov and Pinchuk, 2015).

3.6 CONCLUSION

Domestic natural gas demand has recently been on the decline. However, in light of the greater decreases in domestic natural gas production, particularly from the EU's two greatest producers, the Netherlands and the United Kingdom, import dependency has increased. Though Norway is the second largest natural gas exporter to the EU, Russia accounts for the majority of EU natural gas imports and is likely to play a significant role in future EU natural gas. Member State dependency on Russian imports varies but some find themselves completely dependent on these imports.

Gas relations between the EU and Russia date back to the final years of the 1960s when both gas markets consisted of national vertically integrated monopolies. However, as the EU unbundled its natural gas sector, the once harmonious industries diverged. Natural gas is vital to the Russian economy with the European market serving as Russia's main client. For this reason, Russia relies on EU natural gas demand. However, Russia also uses natural gas as a political tool to the dismay of the EU. Particularly in response to the shutting off of gas to Ukraine in 2009, the EU has pushed policies to increase its energy security. This includes efforts to liberalise the gas market including spot pricing instead of TOP contracts or unbundling of energy companies. In contrast to this, however, market interventionist approaches are assessed, such as the idea of a European Energy Union to provide one unified EU voice on energy contract negotiations, as these are currently carried out bilaterally to the advantage and disadvantage of certain Member States. Among other reasons, this is a complex strategy due to highly diverse Member State preferences. Beyond this, the EU is looking to diversify its natural gas supplies by exploring increased LNG imports, building additional pipelines and constructing more natural gas and electricity interconnections. The latter will not only make it easier for natural gas to be transported to where it is needed, but the additional improvements to electricity interconnectors allows renewable energy to better reach demand in other parts of the EU, potentially reducing the need for natural gas for electricity generation.

The introduction of ambitious climate policies seems to be the most promising long-term approach to reduce energy consumption overall and fossil fuel consumption in particular. Renewable energy can substitute high carbon fossil fuels and nuclear power and also significantly reduce natural gas, which is likely to act as a bridge technology. Furthermore, energy efficiency improvements reduce total energy consumption and therefore speed up the effect of energy carrier substitution.

With eyes towards the future, EU natural gas import needs in the decades to come are somewhat uncertain and mainly depend on compliance with existing climate legislation, stringency of future targets and policies, as well as the market development of other energy carriers, such as LNG. Nevertheless, it is very likely that natural gas will still play some role in the EU energy mix in the long term, which will have to be imported to a large share as domestic sources are running out.

Under this uncertainty, Russia has taken steps to build up its exports to the Asian market, though it seems this will likely not be able to fully replace EU gas demand. Russia also needs to invest in current infrastructure as well as the development of new fields in order to keep its current production levels and secure supplies to the EU. In another effort to secure EU supplies, Russia is diversifying its pipeline infrastructure to avoid areas which might create insecurity for transit. On the other hand, some planned EU and Russian pipelines conflict with one another as politics mixes with energy security interests.

The important question moving forward is how the EU can approach its relationship with Russia when it both depends on it as a natural gas supplier in the short to medium term while it simultaneously pursues a long-term strategy of providing low or no natural gas demand for Russia.

REFERENCES

Bellona, August 29, 2012. Russia's Giant Shtokman Gas Field Project Put on Indefinite Hold over Cost Overruns and Failed Agreements. Retrieved from: http://bellona.org/news/fossil-fuels/oil/2012-08-russias-giant-shtokman-gas-field-project-put-on-indefinite-hold-over-cost-over-runs-and-failed-agreements.

Bilgin, M., 2011. Energy security and Russia's gas strategy: the symbiotic relationship between the state and firms. Communist and Post-Communist Studies. 44 (2), 119–127. http://doi.org/10.1016/j.postcomstud.2011.04.002.

Bochkarev, D., August 1, 2013. Liberalisation of Natural Gas Exports to Help Russia Restore Positions on the European and Global Markets. Retrieved from: http://russiancouncil.ru/en/inner/?id_4=2182.

Boussena, S., Locatelli, C., 2013. Energy institutional and organisational changes in EU and Russia: revisiting gas relations. Energy Policy. 55, 180–189. http://doi.org/10.1016/j.enpol.2012.11.052.

BP, 2005. BP Statistical Review of World Energy June 2005. BP. Retrieved from: http://www.bp.com/statisticalreview.

BP, 2007. BP Statistical Review of World Energy June 2007. BP. Retrieved from: http://www.bp.com/statisticalreview.

BP, 2009. BP Statistical Review of World Energy June 2009. BP. Retrieved from: http://www.bp.com/statisticalreview.

BP, 2011. BP Statistical Review of World Energy June 2011. BP. Retrieved from: http://www.bp.com/statisticalreview.

BP, 2012. BP Statistical Review of World Energy June 2012. BP. Retrieved from: http://www.bp.com/statisticalreview.

BP, 2013. BP Statistical Review of World Energy June 2013. BP. Retrieved from: http://www.bp.com/statisticalreview.

BP, 2014. BP Statistical Review of World Energy June 2014. BP. Retrieved from: http://www.bp.com/statisticalreview.

BP, 2015a. BP Statistical Review of World Energy June 2015. Retrieved from: http://www.bp.com/content/dam/bp/pdf/Energy-economics/statistical-review-2015/bp-statistical-review-of-world-energy-2015-full-report.pdf.

BP, 2015b. BP Energy Outlook 2035 EU. Retrieved from: http://www.bp.com/content/dam/bp-country/de_de/PDFs/Sonstiges/Energy_Outlook_Energietrends_und_Daten_EU_2015.pdf.

Bradshaw, M., Bridge, G., Bouzarovski, S., Watson, J., Dutton, J., 2014. The UK's Global Gas Challenge (Research Report). UK Energy Research Center, London, UK. Retrieved from: http://www.google.de/url?sa=t&rct=j&q=&esrc=s&source=web&cd=2&ved=0CCwQFjAB&url=http%3A%2F%2Fwww.ukerc.ac.uk%2Fasset%2FF4F59C91-7824-4BEF-88FB1A9A11B18528%2F&ei=DaCSVaSeAsmsswHAsoOgCg&usg=AFQjCNFgRPYwrWr8fCECYLj25N2-MlerxQ&bvm=bv.96783405,d.bGg.

van Breevoort, P., Hagemann, M., Höhne, N., Day, T., de Vos, R., 2014. Increasing the EU's Energy Independence Technical Report (No. CLIDE15208) Ecofys. Retrieved from: http://www.ecofys.com/files/files/ecofys-2014-increasing-eu-s-energy-independence-technical-report.pdf.

Buchan, D., 2014. Europe's Energy Security – Caught Between Short-Term Needs and Long-Term Goals. The Oxford Institute for Energy Studies. Retrieved from: http://www.oxfordenergy.org/wpcms/wp-content/uploads/2014/07/Europes-energy-security-caught-between-short-term-needs-and-long-term-goals.pdf.

Chyong, C.-K., Tcherneva, V., March 17, 2015. Europe's Vulnerability on Russian Gas. Retrieved from: http://www.ecfr.eu/article/commentary_europes_vulnerability_on_russian_gas.

Cottier, T., Matteotti-Berkutova, S., Nartova, O., 2010. Third Country Relations in EU Unbundling of Natural Gas Markets: The "Gazprom Clause" of Directive 2009/73 EC and WTO Law (Working Paper No. No 2010/06). NCCR Trade Regulation. Retrieved from: http://www.nccr-trade.org/fileadmin/user_upload/nccr-trade.ch/wp5/Access%20to%20gasgrids.pdf.

Cutler, R., February 23, 2015. Turkmenistan Pipeline From East to West to Be Completed This Year. Retrieved from: http://www.eurasiansecurity.com/energy-geopolitics/turkmenistan-pipeline-completing-east-west/.

Dickel, R., Hassanzadeh, E., Henderson, J., Honoré, A., El-Katiri, L., Pirani, S., Rogers, H., Stern, J., Yafimava, K., 2014. Reducing European Dependence on Russian Gas: Distinguishing Natural Gas Security From Geopolitics. The Oxford Institute for Energy Studies, Oxford, UK. Retrieved from: http://www.oxfordenergy.org/wpcms/wp-content/uploads/2014/10/NG-92.pdf.

EU Commission, 2011a. Energy Roadmap 2050 (No. COM(2011) 885 final) EU Commission, Brussels, Belgium. Retrieved from: http://eur-lex.europa.eu/legal-content/EN/TXT/PDF/?uri= CELEX:52011DC0885&from=EN.

EU Commission, 2011b. Energy Roadmap 2050 Impact Assessment Part 2 Including Part II of Annex 1 "Scenarios – Assumptions and Results" and Annex 2 "Report on Stakeholders Scenarios" (Commission Staff Working Paper No. SEC(2011) 1565) EU Commission, Brussels, Belgium. Retrieved from: https://ec.europa.eu/energy/sites/ener/files/documents/sec_2011_1565_part2. pdf.

EU Commission, 2015. Energy Union Package: Achieving the 10% Electricity Interconnection Target. EU Commission, Brussels, Belgium. Retrieved from: http://ec.europa.eu/priorities/energy-union/docs/interconnectors_en.pdf.

EurActiv, February 13, 2015. Russian Gas Price Seen Falling by up to 35% in Europe. Retrieved from: http://www.euractiv.com/sections/energy/russian-gas-price-seen-falling-35-europe-312104.

EUR-Lex, 1997. Agreement on Partnership and Cooperation. Retrieved from: http://www. russianmission.eu/userfiles/file/partnership_and_cooperation_agreement_1997_english.pdf.

EUR-Lex, January 19, 2011. Security of Supply of Natural Gas. Retrieved from: http://eur-lex. europa.eu/legal-content/EN/TXT/HTML/?uri=URISERV:en0026&from=EN.

European Commission, 2014b. 2030 Framework for Climate and Energy Policies. European Commission. Retrieved from: http://ec.europa.eu/clima/policies/2030/index_en.htm.

European Commission, 2014a. Questions and Answers on Security of Energy Supply in the EU. EU Commission, Brussels, Belgium. Retrieved from: http://europa.eu/rapid/press-release_MEMO-14-379_en.htm.

European Commission, 2015a. Energy Union Package, COM (2015) 80 final (2015). Retrieved from: http://ec.europa.eu/priorities/energy-union/docs/energyunion_en.pdf.

European Commission, July 10, 2015b. Energy: Central Eastern and South Eastern European Countries Join Forces to Create an Integrated Gas Market. Retrieved from: http://europa.eu/rapid/press-release_IP-15-5343_en.htm.

European Commission, August 6, 2015c. LNG and Gas Storage in the EU: Share Your Views! Retrieved from: https://ec.europa.eu/energy/en/news/lng-and-gas-storage-eu-share-your-views.

European Commission, October 8, 2015d. Consultation on an EU Strategy for Liquefied Natural Gas and Gas Storage. Retrieved from: https://ec.europa.eu/energy/en/consultations/consultation-eu-strategy-liquefied-natural-gas-and-gas-storage.

European Commission, October 8, 2015e. Russia. Retrieved from: https://ec.europa.eu/energy/en/topics/international-cooperation/russia.

European Energy Charter, 2007. Retrieved from: http://europa.eu/legislation_summaries/energy/external_dimension_enlargement/l27028_en.htm.

European Union External Action, 2015. EU Relations with Russia. Retrieved from: http://eeas.europa.eu/russia/about/index_en.htm.

Eurostat, 2015a. Smarter, Greener, More Inclusive? Indicators to Support the Europe 2020 Strategy, 2015th ed. Eurostat Statistical books. Retrieved from: http://ec.europa.eu/eurostat/documents/3217494/6655013/KS-EZ-14-001-EN-N.pdf.

Eurostat, March 2, 2015b. Energy Intensity of the Economy: Eurostat Code Tsdec360. Retrieved from: http://ec.europa.eu/eurostat/tgm/table.do?tab=table&init=1&language=en&pcode=tsdec360&plugin=1.

Eurostat, March 2, 2015c. Primary Energy Consumption: Eurostat Code T2020_33. Retrieved from: http://ec.europa.eu/eurostat/tgm/table.do?tab=table&init=1&language=en&pcode=t2020_33&plugin=1.

Eurostat statistical books, 2015. Energy Balance Sheets 2013 Data (No. ISSN 1830–7558). Eurostat. Retrieved from: http://ec.europa.eu/eurostat/documents/3217494/6898731/KS-EN-15-001-EN-N.pdf/e5851c73-9259-462e-befc-6d037dc8216a.

FERC, 2016. North American LNG Import/Export Terminals. Retrieved from: https://www.ferc.gov/industries/gas/indus-act/lng/lng-approved.pdf.

Gas Infrastructure Europe, October 7, 2015. EU 28 Storage Data. Retrieved from: http://transparency.gie.eu/.

Gazprom, 2014a. The Power of Growth. Retrieved from: http://www.gazprom.com/f/posts/55/477129/gazprom-in-figures-2010-2014-en.pdf.

Gazprom, April 22, 2014b. Setting Fair Gas Prices in Russia to Boost Domestic Economy. Retrieved from: http://www.gazprom.com/press/news/2014/april/article189315/.

Gazprom, 2015a. About Gazprom. Retrieved from: http://www.gazprom.com/about/.

Gazprom, 2015b. Eastern Gas Program. Retrieved from: http://www.gazprom.com/about/production/projects/east-program/.

Gazprom, 2015c. Shares. Gazprom. Retrieved from: http://www.gazprom.com/investors/stock/.

Goldthau, A., 2013. The Geopolitics of Natural Gas: The Politics of Natural Gas Development in the European Union. Harvard University's Belfer Center and Rice University's Baker Institute Center for Energy Studies. Retrieved from: http://belfercenter.ksg.harvard.edu/files/MO-CES-pub-GeoGasEU-102513.pdf.

Goldthau, A., Boersma, T., 2014. The 2014 Ukraine-Russia crisis: implications for energy markets and scholarship. Energy Research & Social Science 3, 13–15.

Götz, R., 2014. Gazproms Preispolitik in Europa. Energiewirtschaftliche Tagesfragen. Retrieved from: http://www.academia.edu/9826439/Gazproms_Preispolitik_in_Europa.

Growitsch, C., Stronzik, M., 2014. Ownership Unbundling of Gas Transmission Networks – Empirical Evidence (EWI Working Paper No. No 11/7). Institute of Energy Economics at the University of Cologne (EWI), Cologne, Germany. Retrieved from: http://www.ewi.uni-koeln.de/fileadmin/user_upload/Publikationen/Working_Paper/EWI_WP_11-07_unbundling_gas.pdf.

Hafizoglu, R., October 15, 2015. Nord Stream-2 Reduces Chances of Turkish Stream. Retrieved from: http://en.trend.az/business/economy/2444442.html.

Hafner, M., 2012. Russian Strategy on Infrastructure and Gas Flows in Europe (No. Working Paper N. 73). POLINARES. Retrieved from: http://www.polinares.eu/docs/d5-1/polinares_wp5_chapter5_2.pdf.

Henderson, J., Pirani, S., (Eds.) 2014. The Russian Gas Matrix: How markets are driving change, Oxford: OIES/OUP.

Holz, F., Richter, P.M., Egging, R., 2013. The Role of Natural Gas in a Low-Carbon Europe: Infrastructure and Regional Supply Security in the Global Gas Model. Deutsches Institut für Wirtschaftsforschung, Berlin, Germany. Retrieved from: http://www.diw.de/documents/publikationen/73/diw_01.c.417156.de/dp1273.pdf.

Honoré, A., 2014. The Outlook for Natural Gas Demand in Europe. The Oxford Institute for Energy Studies. Retrieved from: https://www.oxfordenergy.org/2014/06/the-outlook-for-natural-gas-demand-in-europe-2/.

Interfax, May 5, 2014a. Russia Files Suit in WTO against EU Third Energy Package.

Interfax, November 3, 2014b. Gazprom Paves Way for South Stream With Discounts.

International Energy Agency, 2015. Gas Medium-Term Market Report 2015. Retrieved from: http://www.iea.org/Textbase/npsum/MTGMR2015SUM.pdf.

International Gas Union, 2014. World LNG Report – 2014 Edition. Total Retrieved from: http://www.igu.org/sites/default/files/node-page-field_file/IGU%20-%20World%20LNG%20Report%20-%202014%20Edition.pdf.

Jones, D., Gaventa, J., Dufour, M., 2015. Europe's Declining Gas Demand: Implications for Infrastructure Investment and Energy Security. E3G. Retrieved from: http://e3g.org/docs/E3G_Europes_Declining_Gas_Demand_10_6_2015.pdf.

Kropatcheva, E., 2014. He who has the pipeline calls the tune? Russia's energy power against the background of the shale "revolutions.". Energy Policy. 66, 1–10. http://doi.org/10.1016/j.enpol.2013.10.058.

Kuzemko, C., 2014. Ideas, power and change: explaining EU-Russia energy relations. Journal of European Public Policy 21 (1), 58–75.

Lazarus, M., Tempest, K., Klevnäs, P., Korsbakken, J.I., 2015. Natural Gas: Guardrails for a Potential Climate Bridge. Stockholm Environment Institute Retrieved from: http://www.sei-international.org/mediamanager/documents/Publications/Climate/NCE-SEI-2015-Naturalgas-guardrails-climate-bridge.pdf.

Mareš, M., Laryš, M., 2012. Oil and natural gas in Russia's eastern energy strategy: dream or reality? Energy Policy 50, 436–448 (Special Section: Past and Prospective Energy Transitions - Insights from History).

Müller-Kraenner, S., 2008. Energy Security – Re-measuring the World. Earthscan Publications, London.

Natural Gas Europe, 2014. Falling Short: A Reality Check for Global LNG Exports. Retrieved from: http://www.naturalgaseurope.com/global-lng-exports-leonardo-maugeri.

Natural Gas Europe, May 23, 2015a. Russia Carries on with Turkish Stream Pipeline. Retrieved from: http://www.naturalgaseurope.com/russia-barrels-ahead-with-turkish-stream-pipeline-23858.

Natural Gas Europe, June 23, 2015b. EU-Central Asia Energy Dimension: New Positive Steps for a Trans-caspian Corridor? Retrieved from: http://www.naturalgaseurope.com/eu-central-asia-energy-dimension-new-positive-steps-for-a-trans-caspian-corridor-24200.

Natural Gas Europe, July 7, 2015c. Virtual Pipelines. Retrieved from: http://www.naturalgaseurope.com/trans-adriatic-pipeline-and-turkish-stream-repeating-history-24524.

Natural Gas Europe, July 20, 2015d. The EU's Non Energy Union. Retrieved from: http://www.naturalgaseurope.com/eu-non-energy-union-24384.

Nowak, Z., Ćwiek-Karpowicz, J., Godzimirski, J., March 25, 2015. Russia's Grand Gas Strategy – the Power to Dominate Europe? Retrieved from: http://www.energypost.eu/russias-grand-gas-strategy-power-dominate-europe/.

Orttung, R.W., Overland, I., 2011. A limited toolbox: explaining the constraints on Russia's foreign energy policy. Journal of Eurasian Studies 2 (1), 74–85.

Pirani, S., Stern, J., Yafimava, K., 2009. The Russo-Ukrainian Gas Dispute of January 2009: A Comprehensive Assessment. Oxford Institute for Energy Studies. Retrieved from: http://www.oxfordenergy.org/wpcms/wp-content/uploads/2010/11/NG27-TheRussoUkrainianGasDisputeof-January2009AComprehensiveAssessment-JonathanSternSimonPiraniKatjaYafimava-2009.pdf.

Rogozhin, A., May 24, 2015. Will American LNG help Europe Out ? New Eastern Outlook. Retrieved from: http://journal-neo.org/2015/05/24/rus-pomozhet-li-evrope-amerikanskij-spg/.

RT, May 2014. China to Pay about $390 for Russian Gas – Energy Expert Tells RT. Retrieved from: http://rt.com/business/162016-china-russia-gas-price/.

Söderbergh, B., Jakobsson, K., Aleklett, K., 2010. European Energy Security: An Analysis of Future Russian Natural Gas Production and Exports (Energy Policy) Elsevier, Uppsala, Sweden. Retrieved from: http://www.sciencedirect.com/science/article/pii/S0301421510006579.

Stoppard, M., 2015. Low Oil Prices and LNG. IHS Energy. Retrieved from: http://ceraweek.com/2015/wp-content/uploads/2015/04/WSJ2015-04-22final.pdf.

The Economist Intelligence Unit, July 9, 2014. Coal's Last Gasp in Europe. The Economist. Retrieved from: http://www.eiu.com/industry/article/741997658/coals-last-gasp-in-europe/2014-07-09.

The Moscow Times, December 11, 2014. Russia's Gazprom Wants to Raise Domestic Gas Prices as Ruble Plunge Hits Margins. Retrieved from: http://www.themoscowtimes.com/business/article/russias-gazprom-wants-to-raise-domestic-gas-prices-as-ruble-plunge-hits-margins/513181.html.

The unwelcome renaissance, January 5, 2013. The Economist. Retrieved from: http://www.economist.com/news/briefing/21569039-europes-energy-policy-delivers-worst-all-possible-worlds-unwelcome-renaissance.

Trans Adriatic Pipeline, 2015. Southern Gas Corridor. Retrieved from: http://www.tap-ag.com/the-pipeline/the-big-picture/southern-gas-corridor.

Talseth, L.-C.U., 2012. The EU-Russia Energy Dialogue: Travelling Without Moving. German Institute for International and Security Affairs, Berlin, Germany. Retrieved from: http://www.swp-berlin.org/fileadmin/contents/products/arbeitspapiere/talseth_20120402_KS.pdf.

de Vos, R., van Breevoort, P., Hagemann, M., Höhne, N., 2014. Increasing the EU's Energy Independence: A No-regrets Strategy for Energy Security and Climate Change. Open Climate Network, Ecofys. Retrieved from: http://www.ecofys.com/files/files/ecofys-ocn-2014-increasing-the-eus-energy-independence.pdf.

World Coal Association, 2012. Coal Market & Transportation. Retrieved July 31, 2015 from: http://www.worldcoal.org/coal/market-amp-transportation/.

World Trade Organisation, July 20, 2015. European Union and its Member States – Certain Measures Relating to the Energy Sector. Retrieved from: https://www.wto.org/english/tratop_e/dispu_e/cases_e/ds476_e.htm.

YCharts, June 2015a. Japan Liquified Natural Gas Import Price. Retrieved from: https://ycharts.com/indicators/japan_liquefied_natural_gas_import_price.

YCharts, August 2015b. Average Crude Oil Spot Price. Retrieved from: https://ycharts.com/indicators/average_crude_oil_spot_price.

YCharts, August 2015c. European Union Natural Gas Import Price. Retrieved from: https://ycharts.com/indicators/europe_natural_gas_price.

Zhdannikov, D., Pinchuk, D., June 18, 2015. UPDATE 2-Russia's Gazprom to Expand Nord Stream Gas Pipeline with E.on, Shell. OMV. Retrieved from: http://uk.reuters.com/article/2015/06/18/energy-gazprom-pipeline-idUKL5N0Z42OB20150618.

Chapter 4

The Macroregional Geopolitics of Energy Security: Towards a New Energy World Order?

G. Cotella, S. Crivello

Interuniversity Department of Regional and Urban Studies and Planning (DIST), Politecnico di Torino, Turin, Italy

4.1 INTRODUCTION

Energy sources are essential to the economic activity that sustains and improves the quality of life. As a consequence to this statement, energy plays a crucial role in international affairs (Klare, 2008a; Bradshaw, 2009; Favennec, 2011) and international relations with energy suppliers and issues of energy security are central nodes of geopolitical tensions and political agendas of countries all over the world (Goodstein, 2004; Roberts, 2004) and of the European Union (EU) (Youngs, 2009; Umbach, 2010; Bosse, 2011). A major cause of this situation is the mismatch between the location of resources and demand that leads to energy products representing the largest traded commodities worldwide. This poses a global challenge since the largest energy consuming economies (among which the EU, the United States and China) lack adequate indigenous resources to support their energy needs. Most of the energy resources are indeed concentrated in politically unstable areas such as the Middle East and, to a lesser extent, in other regions like Africa, Latin America, Russia and the Caspian Sea.

Such a situation is not new: historically, industrialised countries started expressing concerns over energy security mainly in terms of secure energy supply as early as in the first part of the 20th century. The so-called British fuel switch decision[1] inextricably connected the security of oil supply with the conduct and preparation of several war events during World War I and II, thereby putting forward the crucial role of energy as a strategic asset in foreign policy

---

1. Winston Churchill's decision in August 1911 that the British Royal Navy needed to convert from the easily accessible and politically secure 'Welsh Coal' to the volatile 'Persian Oil' in order to maintain its military dominance was the first signal of the growing intensity of global competition over energy sources.

Low-carbon Energy Security from a European Perspective. http://dx.doi.org/10.1016/B978-0-12-802970-1.00004-8
Copyright © 2016 Elsevier Ltd. All rights reserved.

and military conflicts. In spite of that, the availability of relatively cheap energy resources also played a major role in the reconstruction and development of Europe and Japan in the aftermath of World War II. This prolonged era of relative confidence in the availability of abundant and secure energy resources came to an abrupt end following the outbreak of the 1973 Arab–Israeli War.

This rapid overview highlights the evident geopolitical implications in terms of the interconnections between energy security and energy transition and, in turn, of main fuel switches. At the beginning of the 20th century, there has been a switch from coal – the fuel that propelled the Industrial Revolution in Great Britain and the world in the 19th century – to oil. After the oil crises, in the 1970s, there has been the gradual replacement of oil by natural gas, with differences in market dynamics (from world to regional markets). Then, the growing understanding of the challenges posed by climate change added a new dimension to the energy security concept, no longer limited to availability of energy resources at affordable prices but with environmental considerations in it.[2] Increasing concerns about environmental protection and climate change also encouraged a transition from oil and natural gas to low-carbon energy sources, mostly in terms of renewable energy sources and with the controversial role of nuclear power.

In short, the energy system in the last century appears to be moving towards a decarbonisation and diversification of the energy portfolio with an increasing weight of low-carbon and locally available energy sources. This can be related to two major issues. On the one hand, the need for lessening one country's overall dependence on a single energy source because of limited physical reserves, geopolitical risks, etc.; on the other hand, the rising environmental concerns about the sustainability of the current development and growth model. Aware of these issues, the present chapter focuses on the macroregional geopolitics of energy security,[3] aiming as it is at picturing the future energy geopolitics of Europe, to be read within the broader energy world order. To do so, this chapter is articulated in four main parts. It starts from a discussion of the main storylines that characterise the international debate concerning energy security (eg, the global energy consumption growth, the depletion of energy resources, the role of energy technology, etc.), to then move to explore the potential futures of the energy world order. The role played by oil and natural gas and the geopolitical relations caused by the latter will be explored, together with the regions playing a key role in current geopolitical energy situation (ie, the Persian Gulf, the Caspian Sea and Africa).

---

2. See Chapter 8 for an extensive discussion of the meaning of Energy security and the links entwining the concepts of energy security and climate change.

3. The text refers to 'macroregional geopolitics' as the practice of analysing actually existing or potential energy conflicts, spatial relations and functional interactions based on the logics of energy production, consumption or supply that Europe (or relevant parts of Europe) is developing or may develop in the future with external territories such as Russia, the Caspian Sea area and the Northern Africa region. Obviously, the analysis of the EU's energy relations implies the consideration of a large space and a global perspective. European energy relations, in fact, spread above a large part of Asia and Africa, and to a minor degree even the Americas.

Lastly, the chapter proposes an evaluation of geopolitical tensions between different areas in the world, combining and representing the information, perspectives and key nodes for EU energy security through qualitative visual representations.

4.2 ENERGY SECURITY IN THE INTERNATIONAL DEBATE

As already mentioned in various occasions throughout the volume, energy security is a complex issue involving political, economic, social-cultural, technological and environmental dimensions as well as various considerable threats (Vivoda, 2010; Winzer, 2012). Therefore, the analysis of its macroregional geopolitics requires a systemic and integrated approach based on an assessment of multidimensional elements, regarding political, techno-economic and sociocultural aspects of the energy system. The application of this approach also requires a better understanding of historical as well as contemporary debates on energy transition.

As a matter of fact, over the past years energy challenges and environmental concerns broadly have been discussed by the G8 Leaders, the North Atlantic Treaty Organisation (NATO), the United Nations as well as by a plethora of energy influential organisations.[4] Moreover, apart from international organisations debates, there is growing number of literature widely researching energy trends, scenarios as well as climate stabilisation and low-carbon transitions, geopolitical and macroeconomic challenges, and technological, societal and environmental perspectives. As it will be briefly shown in the following subsections, the day-to-day energy and climate change issues have become the focus of increasing international attention for a number of reasons.

4.2.1 The Global Energy Consumption Growth and the Depletion of Energy Resources

First of all, the global energy consumption grows quickly: the global fuel consumption was increasing from 280,622 quadrillion Btu in 1980 to almost double of the energy consumption which had reached 506,853 quadrillion Btu in 2010. Fossil fuels such as oil (34.5%), coal (29%) and natural gas (23.1%) continue to dominate as the main sources of energy produced and consumed worldwide. The shares of nuclear energy and hydroelectricity are only 5.4% and 6.5%, respectively, but the share of renewables while increasing from 0.1% in 1980 to 1.5% in 2010 is still limited (EIA, 2013).

Furthermore, available long-term scenarios from international organisations show that the global energy consumption will continue to grow due to the economic development, urbanisation and population increase (IPCC, 2000, 2008; UNDP, 2001; UNEP, 2006; IEA, 2004). According to the International

---

4. Such as World Energy Council (WEC), World Petroleum Organisation (WPO), Greenpeace, World Council for Renewable Energy (WCRE), International Renewable Energy Agency (IRENA), International Research Network for Low-carbon Societies (LCS-Rnet) and others.

Energy Agency (IEA), over the next 25 years, the world population is projected to grow to almost 9 billion people. The global energy demand will increase approximately 100% over the period 2010–2050.

Within this dramatic scenario, the EU is one of the largest energy consuming regions in the world. Over recent decades, energy consumption in European countries (EU27) rose from 71,747 in 1980 to 83,824 quadrillion Btu in 2010 with petroleum and gas contribution 31,505 quadrillion Btu (18% of world's oil consumption) and 20,935 (17.9% of world's gas consumption) and total energy consumption is forecast to increase by 20% by 2030.

When coming to the availability of energy resources, on the other hand, the world reserves of primary energy and raw materials are, obviously, limited in size. According to recent estimates, the reserves will last another 218 years for coal, 41 years for oil and 63 years for natural gas, under a business-as-usual scenario (EIA, 2013). The petroleum age began about 150 years ago. Easily available energy has supported major advances in agriculture, industry, transportation, and indeed many diverse activities valued by humans. Now world petroleum and natural gas supplies have peaked and their supplies will slowly decline over the next 40–45 years until depleted. Although small amounts of petroleum and natural gas will remain underground, it will be energetically and economically impossible to extract them.

4.2.2 Energy Security Threats

Second, a major topic on many political agendas is security of supply. This challenge has risen in importance on the international policy agenda due to growing dependence of industrialised economies on imported energy consumption and the increased frequency of disruptions in supply. For instance, Europe currently imports approximately 55% of its energy and might reach 70% energy import in the next 20–30 years. It is expected in 2030 that Europe will be importing 85% of its gas, 60% of its coal and 95% of its oil (EC, 2006). Most of the imported energy resources come from countries with geopolitical risks, unstable or potentially unstable political, economic and social situations. Furthermore around 75% of the world's proven oil reserves are located in eight countries (Saudi Arabia, Venezuela, Canada, Iran, Iraq, Kuwait, United Arab Emirates and Russia) and almost half of the natural gas in three of them (Russia, Iran and Turkmenistan) (EIA, 2013).

Additionally, NATO defines geopolitical risks with potential government decisions to suspend deliveries because of deliberate policies, war, civil strife and terrorism. Energy industries in supplier countries are subject to extensive government interference, and do not function in a competitive market framework. For instance, some nations have prohibited foreign investment in their energy sectors, while others have demanded a greater share of control or revenues (eg, Russia, Venezuela, Iran, Turkmenistan and Kazakhstan). NATO also stresses that energy will increasingly be used as a political weapon. A good example of this is the Russian–Ukrainian conflict, which shows how energy rich countries can force other poor energy countries. In addition, security of supply is threatened by political instability of exporting regions where civil wars, local conflicts and terrorism have often been cause of temporary damage of energy facilities and infrastructures (Laurmann, 1992; Bentley, 2002; Correljé and van der Linde, 2006; Bilgin, 2009; Toft et al., 2010).

4.2.3 Environmental Pressure and Long-Term Targets

Third, using fossil fuels still has a massive impact on the environment and the climate. The Intergovernmental Panel on Climate Change (IPCC) report indicates that the 20th century saw a considerable and sudden increase in global temperature when compared to the last 1000 years (IPCC, 2000). The measured temperature rise is partly attributed to the emission of greenhouse gases (GHG) from fossil fuels combustion (Soytas and Sari, 2009; Acaravci and Ozturk, 2010; Pao and Tsai, 2010; Lean and Smyth, 2010; Apergis and Payne, 2010; Arouri et al., 2012). The emission of carbon dioxide gas (CO_2) has increased significantly and this negative trend continues increasing every year since 1900. The global CO_2 emission has risen from 530 million tonnes in 1900 to 8700 million tonnes in 2010 with contribution of China (share 29%), the United States (16%), the EU (EU27) (11%), India (6%) and the Russian Federation (5%), followed by Japan (4%). According to IPCC Special Report on Emissions Scenarios (SRES) with current climate change mitigation policies and related sustainable development practices, global GHG emissions will continue to grow over the next few decades. IPCC SRES scenarios predict a 25–90% increase of GHG emissions in 2030 relative to 2000.

Beyond increasing GHG emission levels, widespread impacts are expected on food and water supplies, weather patterns, ecosystem stability and, of course, energy production itself. Exceptional natural disasters could delay the exploration of oil and natural gas fields, and more extreme weather situations could force the shutdown of oil production in the affected regions. Preventing that harm will require a fundamental transformation of the energy system. Clearly energy transition changes are likely to be expensive. However, studies have concluded that ignoring the problem will cost even more. The Stern Review, for example, concluded that, in case of inaction, the overall costs and risks of climate change will be equivalent to losing at least 5% of global GDP each year (Stern Review, 2006). The same study estimates the cost of GHG emissions mitigation to be between 1% and 2% of GDP per year. Thus in the UN Climate Change Conference, it was stressed that it is mandatory to put in place a robust policy mechanism to achieve the stabilisation of GHGs in the atmosphere.

4.2.4 Energy Policies and Beyond

As a shown above, the stable energy supply and climate change are clearly one of the most difficult challenges faced by the energy sector, now, and for the decades to come.

As a mentioned before, the world continue to consume more energy as well as produce more CO_2 emissions, at the same time geopolitical situation around energy producing countries keeps tense. It is, obviously, energy efficiency and renewable energy technologies as well as change in consumer behaviour will play a very important role in reaching low-carbon societies. And as any measures there are urgent needs to create energy policy. Many countries have already implemented their energy or low-carbon policy. Despite these initiatives, current efforts are still not enough to reach low-carbon societies. Many new long-term scenarios show that new and more radical energy and climate policies are needed to avoid humanity becoming locked into carbon-intensive development paths (IPCC, 2007; EIA, 2013; see also Laconte, 2011).

At the same time energy policies need to be placed in a broader context encompassing urban, transportation, agricultural, technological and societal policies. For instance, cities are hubs of innovation and creativity and will be central to achieving a sustainable future. An increasing number of initiatives limiting GHG emissions in cities in industrialised countries are already in place. These initiatives include the deployment of low-carbon mass transportation systems such as bus rapid transit systems, car-sharing arrangements, cycling infrastructure, support for public transport, solar thermal technologies for hot water supply, energy efficiency projects for public buildings and integrated waste management.

4.3 THE MACROREGIONAL GEOPOLITICS OF ENERGY SECURITY

As mentioned above, the availability of secure energy sources is today essential for the proper functioning of any economy (Klare, 2008a). Due to this reason, competition for energy is extreme today (Peters, 2004; Klare, 2008b). Whereas in the aftermaths of World War II, core industrial countries such as the United States, the United Kingdom and Japan accounted for a large share of global energy consumption, today a number of 'new' emerging countries are driving further increases in the demand for energy sources (Klare, 2008a). According to the EU-funded POLINARES project,[5] new actors such as China, India and Brazil now play an important role on the international stage both as engines of demand and also leading producers of minerals and energy resources. Russia and other countries, which emerged from the break-up of the Soviet Union, are also significant forces in the oil, gas and mineral markets. However, a pivotal role is played by China (Cornelius and Story, 2007). In 1990, China accounted for 8% of global energy consumption, while the United States accounted for 24% and Europe for 20%. With the growth

5. POLINARES – *The Changing Oil Value Chain: Implications for Security of Supply* (www.polinares.eu) is an EU-funded research project exploring global challenges in the competition for natural resources and proposing new approaches to collaborative solutions.

of Chinese economy, the situation has changed radically: in 2010, China surpassed the United States becoming the most important country in terms of energy consumption. It is easy to figure that China will find more and more difficult, in the future, to get further energy supplies. Chinese policy makers will probably try to raise both local energy production and control over external energy sources (Li, 2003).

One should notice how most of the energy used in the world is still provided by fossil fuels (oil, coal and natural gas): oil is the predominant source (33% of total energy consumption), followed by coal (27%) and natural gas (24%). Renewable sources, with an average annual growth of 1.8% since 1990, currently provide about 13% of global energy consumption. Nuclear energy provides about 6% (EIA, 2013). A growing number of facts and figures suggest that the 'easy oil' era will be replaced by a 'difficult oil' era (Roberts, 2004). The marginal cost of oil production, despite being under a prolonged period of relatively low price for oil, is expected to increase: every new oil barrel added to the global reserve will be more difficult and more costly than the previous one, according to scholars. In fact, every new barrel will be extracted deeper in the ocean, in less accessible places and in dangerous spaces, for example, because of possible wars (Roberts, 2004; Jojarth, 2008) and similar scenarios that will most probably characterise all other energy sources, such as carbon, natural gas and uranium (Goldthau and Witte, 2010).

Moreover, it should be stressed that the world is changing from a regime characterised by liberal market principles to one in which state capitalism is more prevalent than in the 1990s. It is widely recognised that the world is currently in transition from a political and economic regime in which liberal market values were prevalent, even if not dominant, to one in which State Capitalist values appear to be gaining more adherents. A consequence may be that energy and mineral prices will be volatile, that markets will be fragmented or that partial supply interruptions will occur for some actors, even though there will be no absolute shortage of resources. This transition is occurring at the same time as demand for energy and mineral resources is rising. The result is a greater degree of unpredictability and volatility in international commodity markets.

As mentioned, policy makers and business managers will keep on building strategies based on the idea that fossil fuels will still be the main energy sources for the planet for several decades. In the words of the Energy Information Administration (EIA, 2013), it is expected that in 2030 fossil fuels will provide about 87% of the world's energy needs. According to these scenarios, most countries will still rely on traditional fuels, with a consequent increase in the competition for the control of unexploited energy reserves (Klare, 2001; Bradshaw, 2009). As stated by the POLINARES project, the increasing interdependence of the world's nations in the context of energy and minerals is likely to cause tensions and conflicts that may undermine future global peace and economic development.

4.3.1 Towards a New Energy World Order

Building on the above debate, many scholars currently think that with the issue of energy security becoming more and more crucial, concepts as 'power' and 'influence' in the international system will change their meaning (see, eg, Favennec, 2011). In this sense, the scientific debates use the expression 'new energy world order' (Klare, 2008a, p. 7) as opposed to an 'old' energy world order in which each country occupies a place in a hypothetical hierarchy of States according to its endowment in terms of nuclear missiles, warships and soldiers (Cohen, 1991). Within this 'new' energy world order, the position of a country in the global ranking seems to be increasingly determined by the possession or control of vast oil and natural gas reserves, or by the capability to mobilise money and relations in order to acquire energy resources from the outside. In this sense, energy surplus or deficit has significant and complex geopolitical and economic implications.[6]

The importance of this topic is also reflected in the field of social research, with various projects that examine the global challenges faced with respect to access to oil, gas and mineral resources and propose solutions for the various policy actors by combining theoretical and empirical analyses from a wide range of disciplines as political science, economics, geology, engineering, technology, law and security studies. The results shows, in brief, that countries with energy deficit will be progressively forced to pay higher prices for imported fuels, competing at the same time with each other in order to secure supplies, ie, to acquire energy sources from countries characterised by energy surplus. On the contrary, energy exporting countries will gain more and more from growths in the cost of energy.[7] Dynamics like these are among the main causes of the fortune of countries like Russia, Dubai and Abu Dhabi (see Acuto, 2010).

An additional element that characterises the mentioned 'new energy world' is the difference between democratic areas (such as the EU) and State Capitalist governments (such as China). More in particular, democratic states generally aim at enforcing upon State capitalist governments specific policy criteria and principles to encourage cooperation in the exploitation of natural resources, that do not fit the values and priorities of the latter, in so doing generating tension in political and economic relations.

Within the global energy order, both energy-exporting and energy-importing countries develop strategies in order to improve their position with respect

---

6. Military power still remains crucial, but its relative importance in geopolitics is decreasing when compared to energy. Saudi Arabia, for example, is characterised by a 'weak' army, but the country occupies a central role in international affairs because of its vast oil reserves. Similarly, countries such as Azerbaijan, Kazakhstan, Angola and Sudan have begun to gain influence, despite their limited size (Klare, 2008a).

7. This phenomenon may be grasped by thinking that, in 2008, oil-exporting countries have gained something like $970 billion from the export of oil, a figure that is three times higher than the one of 2002 (Klare, 2008a).

to actual or potential competitors (Helm, 2002; Peters, 2004). This is evident when looking at the construction of networks and formal/informal agreements between exporting countries, institutions and organisations grouping energy importing countries, and hybrid forms of regionalisation gathering both exporting and importing countries (such as the strategic alliance between China and Russia in order to limit the American influence in Asian energy affairs).[8] A clear sign of this reorganisation may be found in the ongoing nationalisation of energy companies and energy resources in many countries.[9] Of course, energy operators in the private sector still play a significant role, as testified by their colossal profits in recent years, but strategic decisions are more and more in the hands of national governments (Behr, 2010). The most striking example of the tendency towards resource nationalism is probably the one of Vladimir Putin, who led the Kremlin towards national control of oil and gas recourses and who transformed Gazprom, the Russian national enterprise with a monopolistic position in the field of natural gas, in one of the richest and most powerful energy companies of the world (Stern, 2005; Champion, 2006). Also, the case of Japan, a country characterised by a huge energy deficit, testifies to the tendency towards resource nationalism. In fact, Japan supports national energy companies in the seek for secure oil supplies overseas (Hisane, 2006). On their hand some European countries as France and Italy have pursued a different strategy, promoting the development of strategic connection with energy-exporting countries, particularly in Africa, where it is possible to take advantage of sociocultural and economic networks with former colonies (Klare, 2008a).

This 'resource nationalism' is so diffused that it may be conceptualised as a phenomenon echoing the old 'arms race'. Control over oil, natural gas and other energy resources is considered crucial, and in this sense geopolitical relations are evolving according to the logics of energy security (Behr, 2010; Bradshaw, 2009; Favennec, 2011; Goldthau and Witte, 2010).

4.3.2 The Role of Oil in the World and in Europe

Despite global energy policies that are aiming at reducing oil consumption and promoting the differentiation of energy sources (EIA, 2013), oil is still the main energy source in the world (33% of the total energy consumption in 2012). In Europe oil is the main fuel, accounting for 35% of energy consumption (compared

---

8. Although it is too early to predict the overall impact of these agreements, many scholars believe in an ongoing, radical realignment of political powers in order to secure the exploitation of energy resources (Klare, 2004; Bradshaw, 2009; Bosse, 2011).

9. Until recently, most of the world's oil reserves were controlled by large Western private companies (like Exxon Mobil, Chevron, British Petroleum, Royal Dutch Shell, Total SA, etc.). Today, national oil companies control more than 80% of the known oil reserves. Giant players like Saudi Aramco (Saudi Arabia), National Iranian Oil Company (Iran), Petroleos de Venezuela SA (Venezuela) and Gazprom (Russia) play a crucial strategic role in economic and geopolitical terms. In all these cases, the companies are wholly or largely owned by local governments (Klare, 2008a).

to 24% of natural gas; see Chapter 2; see also BP, 2013). In the 1950s, with the economic boom of Western countries, about 2000 billion barrels of crude oil have been produced (source: BP, 2013). New explorations and the constant discovery of new deposits allowed, with time, impressive leaps in terms of oil production (10 million barrels per day in 1950, 25 million in 1962, 50 million in 1971, 75 million at the end of the last century and 86 million in 2012; EIA, 2010; BP, 2013). Particularly, between 1950 and 1970 several giant reserves were discovered in the northern area of Alaska, in the area of the North Sea between the United Kingdom and Norway, and in the Gulf of Guinea in Africa.

However, in the last decades of the 20th century there had been a slowdown in the discovery of new fields and, since the early 1970s, starting with the 1973 energy crisis, concerns about the limits of oil stocks – being a scarce and exhaustible resource – began to rise. Various experts began to question the capability of the energy industry to ensure increases in oil production (see, above all, the famous and controversial 'The Limits to Growth', 1972) and many scholars raised the alarm on the fact that, with similar figures, the 'peak' of the extraction was quickly approaching (Curtis, 2009; Hall and Day, 2009).[10]

Oil is an energy source characterised by high territoriality: it is inextricably linked to the places where crude oil is extracted and where transport infrastructures are located. Extraction sites are geographically concentrated in areas which are not rarely distant from places of consumption, and for this reason oil has to be moved over long distances through different countries. The management of oil security, therefore, involves a number of countries and places, including extraction, transit and consumption sites. According to British Petroleum (BP, 2013), more than 50% of the oil reserves in the world are currently located in the Middle East. The largest shares of reserves are located in Saudi Arabia (15.9%), Iran (9.4%), Iraq (9%), Kuwait (6.1%) and the United Arab Emirates (5.9%). Outside of the Middle East only Venezuela (17.8%), Canada (10.4%) and Russia (5.2%) possess relevant reserves (BP, 2013).

Europe is characterised by high needs of oil: the external demand in 2012 has been of 639 million tonnes (EC, 2013), and oil imports come from a large number of countries. At the same time, European oil production is pretty low, that is about 185 million tonnes in 2012 (EC, 2013). Indeed, as already mentioned in Chapter 2, Europe may be considered a 'single' market for many perspectives, characterised for example by common environmental laws. However, European countries undertake relations with different external suppliers and by different oil transport systems (Fig. 4.1). The EU mainly

---

10. More worryingly, other scholars believe that the peak of oil production has already been reached and that we are right now experiencing a decline in production (Kerr, 2011), showing how the optimism that prevailed at the beginning of the last century has now vanished. On the other hand, it must be mentioned that some countries are apparently distant from their productive peak. This is the case of Canada, Venezuela, Iran and Iraq, whose oil reserves are expected to have a residual life of more than 100 years. It is therefore evident their pivotal strategic role in forthcoming international scenarios (Klare, 2008a).

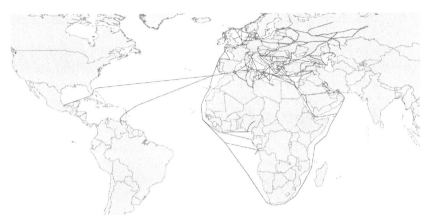

FIGURE 4.1   Main oil corridors in the world to and from Europe. *Reproduced from: Gerboni, R., Grosso, D., Schranz, L., 2014. REACCESS EU Project Outcomes Elaboration. Politecnico of Turin.*

acquires oil from five areas: the North Sea (particularly Norway), Russia, the Caspian area (Kazakhstan and Azerbaijan), Middle East (Saudi Arabia, Iran and Iraq) and North Africa (Libya and Nigeria). Geographical proximity is therefore crucial in the European oil supply scenario, but it has to be mentioned that about 12.8% of European imports come from two distant areas: Venezuela and West Africa (EC, 2013). Offshore deposits in the North Sea belong to Norway, the United Kingdom and Denmark. Oil extracted from the North Sea is for two-thirds destined to Europe, and for one-third to North America. Crude oil refined in Norway, the United Kingdom and Denmark is brought to Europe via undersea pipelines or ships. Supplies from the North Sea are considered highly secure because of the stability of the European market and because of the good relations between European countries (Correlje and Van der Linde, 2006), but actually oil reserves in the North Sea are scarce (15 billion barrels).[11]

Every year, Europe imports about 150 million tonnes of oil from the Middle East. Oil imports from this area used to be higher in the past, but with time – as a consequence of the various oil shocks experienced during the years – European countries have diversified oil suppliers, increasing imports from the North Sea, the Caspian area and Russia. Similarly, the yearly imports from North Africa amount to about 100 million tonnes, particularly from Libya and Algeria. European relations with these two countries were tense in the past, especially in the aftermath of the colonial era. Relations are today controversial and oil supplies can't be taken for granted (Correlje and Van der Linde, 2006; Dabashi, 2012). Oil from the Middle East and North Africa (the MENA region) arrives in Europe mainly through pipelines running through the coasts of Syria, Lebanon, Israel,

---

11. With an average oil production of about 4.5 million barrels per day, the estimated residual life of oil in the North Sea is 10 years (EC, 2013).

Egypt, Libya, Tunisia, Algeria and Morocco. From Morocco, oil is transported by ship, as there are no pipelines crossing the Mediterranean.[12]

Russia and the Caspian Sea countries provide about 330 million tonnes of oil every year. Given current estimated reserves of about 120 billion barrels, production is expected to be assured for at least 30 years. Europe is the main importer for this area, followed by China and United States (each one importing about 20 million tonnes of oil every year according to BP Statistical Review of World Energy, 2013). It has to be mentioned that trade relations between Russia and Europe developed after the collapse of the Soviet Union. In the Soviet Union the extractive and the energy industries used to be highly integrated between Russia and the other countries of the area (Soviet zone of influence). But with the economic crises that characterised the Soviet transition towards a liberal market economy, Russia promoted direct exports to Europe, bypassing the Soviet zone of influence (Champion, 2006). This trend has determined the rise of Russian energy industry but also geopolitical tensions between former Soviet countries, as it is particularly evident in the case of natural gas (Stern, 2005; Aalto, 2008). Oil from Russia and Caucasian area gets in Europe through various corridors, and particularly:

- The Druzhba Pipeline (also known as the 'friendship pipeline' and 'Comecon pipeline') is one of the longest in the world, running through Russia, Ukraine, Hungary, Poland and Germany, with an approximate length of 4000 km. Originally, it was intended to provide Russian oil to satellite Soviet zone of influence. Today, it mainly allows the movement of Russian and Kazakh oil towards Europe.
- The Baltic Pipeline System, a Russian oil transport system operated by the oil pipeline company Transneft. The Baltic Pipeline System transports oil from the Timan-Pechora region, West Siberia and Urals-Volga regions to Primorsk oil terminal at the eastern part of the Gulf of Finland. The pipeline has been completed in 2001 and reached full design capacity in 2006.
- The Sever Pipeline (also known as Kstovo–Yaroslavl–Kirishi–Primorsk Pipeline) is an oil product pipeline in north-west Russia inaugurated in 2008. It transports diesel fuel EN-590. The pipeline is owned and operated by Transnefteproduct, a subsidiary of Transneft. The 1056 km pipeline runs from Kstovo through Yaroslavl and Kirishi to Primorsk, Leningrad Oblast.

As mentioned, under the Soviet Union Caucasian states used to maintain direct connections with Russia, with Moscow that controlled Caspian energy reserves and the pipeline networks were constructed so as to link all the energy-rich countries to Russia. Although Russian hegemony in the area is still visible,

---

12. Current infrastructural projects refer to: (1) two undersea pipelines that will connect Turkey to Syria and Egypt; (2) three pipelines that will connect Italy to Libya, Tunisia and Algeria; (3) two pipelines that are expected to connect Algeria to France and Spain; and (4) one pipeline connecting Morocco with Spain and Portugal.

Caucasian states have partly opened their markets to direct commercial relations with the EU. The Soviet Union's demise, in fact, opened the region to external actors allowing foreign companies to invest in exploiting energy reserves and constructing alternative pipeline routes to transport gas and oil from the region to the international markets. As mentioned, Caspian oil reserves are low when compared to those in the Middle East; what makes Caspian energy resources so significant is that they offer Western buyers the opportunity to diversify energy imports away from the nearly monopolistic energy supplies of the Middle East and Russia.[13]

4.3.3 The Role of Natural Gas in the World and in Europe

Natural gas consumption quickly rose during the last decades: currently, global consumption surpasses 3000 billion m^3 per year, that is, 24% of global energy sources (EIA, 2013). Demand for natural gas is expected to increase in the future, and current reserves – about 180,000 billion m^3 – will assure global supplies for about 60 years (BP, 2013). Natural gas is today a crucial element of the energy mix. Natural gas is employed all over the world for the production of electricity, heating, as a raw material in many industries, and as fuel in the transport sector. The high energy efficiency of natural gas, together with the discussed fears for oil depletion, promotes the use of natural gas in many countries (Victor et al., 2006; Selley, 2013).

The distribution of natural gas extraction sites is even more geographically concentrated than in the case of oil: Iran, Russia and Qatar control about half of the world reserves, while the other eight countries (Turkmenistan, the United States, Saudi Arabia, the United Arab Emirates, Venezuela, Nigeria, Algeria and Australia), as a whole, control a further 21%. With the exceptions of Venezuela, the United States and Australia (controlling together 9.5% of world reserves) all these countries are located in Africa, in the Persian Gulf area and in the former Soviet Union (BP, 2013).

Natural gas presents tight linkages with territorial proximity, even more than oil. Natural gas is too voluminous to be moved by other means than pipelines. As it will be discussed later, the main challenge with natural gas corridors is to maintain relatively constant flows of supplies (Victor et al., 2006; Aalto, 2008). In the case of countries non connected through pipelines – for example because separated by oceans – the only possibility is to import natural gas in liquid form

---

13. Currently, strategic corridors in the Caucasian area are (1) the BTC pipeline running from Baku (Azerbaijan) via Tbilisi (Georgia) to Ceyhan (Turkey); (2) The Baku to Novorossiisk (Russia) pipeline and the Baku to Supsa (Georgia) pipeline. From a geopolitical point of view, oil (and gas) corridors in the area are highly controversial. Russia considers them as political projects challenging Russian security, and Russian political and economic interests. Since Putin's presidency, Russia has emphasised a greater strategic interest in maintaining its influence in the 'near abroad' (Badalyan, 2011). Clearly, redirecting Caspian energy exports away from the Russian transit system challenged not only Russia's dominant role as a key channel for Caspian energy supplies to Europe but also its traditional strategic interests in the Caucasus.

(liquefied natural gas), involving complex and expensive processes of gasification and cooling (Klare, 2008a).

As far as the EU is concerned, consumption of natural gas accounted for 24% in the total energy mix in 2012, a figure fully in line with global trends. Specifically, the EU consumed 520 billion m^3, with an increase of 7.2% with respect to 2011 (EC, 2013). Also in the case of natural gas, Europe strongly depends on external supplies: with the exceptions of Netherlands and Denmark, all the other countries are net importers (EC, 2013). Europe imports every year more than 330 billion m^3 of natural gas via pipelines and 50 billion m^3 in liquid form. About 75% of imported natural gas comes from three countries: Russia, Norway and Algeria. More than 80% of natural gas exported from Russia and Algeria is directed to Europe, as the majority of natural gas comes from Norway.

Natural gas arrives in Europe through three different paths (Fig. 4.2):

- From North Africa (Algeria and Libya) through four pipelines: Transmed (connecting Algeria and Italy through Tunisia), Greenstream (connecting Libya and Italy); Maghreb (connecting Algeria with Spain via Morocco) and Medgas (connecting Algeria with Spanish coasts).
- From Northern Europe through pipelines from the North Sea (Langeled Gas Pipeline) connecting Norway, the United Kingdom and Netherlands. Central Europe is also bypassed by pipelines Tenp and Transitgas, carrying natural gas from Netherlands and from the North Sea to Switzerland and Italy.
- From Russia through a number of routes. Nord Stream pipeline, with a total length of 1.224 km and a carrying capacity of about 27.5 billions of m^3 per year (to be amplified in the future) connects Russia and Germany through the Baltic sea, bypassing Ukraine. Yamal runs from Russia to Germany through Belarus and Poland, with a total length of 4200 km. Gas runs from Russia to Austria, Slovenia and Italy. Finally, Blue Stream carries natural gas to Turkey

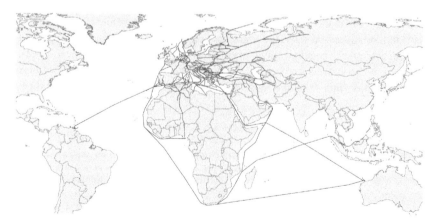

FIGURE 4.2 Main gas corridors in the world to and from Europe. *Reproduced from: Gerboni, R., Grosso, D., Schranz, L., 2014. REACCESS EU Project Outcomes Elaboration. Politecnico of Turin.*

via the Black Sea. Two more pipelines transport natural gas in Turkey from Central Asia: the corridor between Iran and Turkey, and the Baku-Tblisi-Erzurum pipeline.

Recently, the EU is promoting a differentiation of natural gas suppliers in order to reduce energy dependency from Russia. International relations with Russia are, in fact, rather complex,[14] and different from other external suppliers which are considered 'reliable partners' and which agreed with the EU's well-defined economic and contractual frameworks (Youngs, 2009). In Europe, Russia is often considered an 'unreliable' natural gas supplier, particularly because of Russian 'economic menaces' concerning gas exports. On the one hand, European countries are trying to promote alternative ways to acquire natural gas, particularly by developing routes directed to countries other than Russia. On the other hand, Russia has tried to limit the strategic power of transit countries, as Ukraine and Belarus by constructing new pipeline routes (two new sections of North Stream, ended in 2011, and South Stream). North Stream and South Stream may be interpreted as explicit projects aimed at enhancing Russian monopolistic position in the provision on natural gas for Europe (Champion, 2006). With the construction of North Stream, Gazprom will distribute natural gas directly in Germany, Netherlands and in other European countries without interference from Ukraine. Similarly, South Stream (which runs from Russia to Burgas, in Bulgaria, and then to Austria, Italy, Greece, Hungary and Serbia) has decreased the economic feasibility of Nabucco, a pipeline financed by the EU and the United States which is expected to run side by side with South Stream, providing natural gas from Azerbaijan, Iran and Turkmenistan, and not from Russia.

4.3.4 The Role of Coal in the World and in Europe

In the early 1900s coal supplied about 95% of primary energy. Even if the use of oil and gas reduced drastically the share of coal, it remains (with 27% of the total) the second major source (after oil) of primary energy.

According to IEA, coal production has more than doubled since 1980 and coal could replace oil to become again the most important source of energy (IEA, 2013). Thanks to the enormous reserves of coal and thanks to the increasing demand for energy, the use of coal could increase in the future, ensuring security of supply with reduced geopolitical risks. Coal and lignite are, in fact, widely available: proven reserves are sufficient for the next 109 years at current rates of production (BP, 2013). Coal is also widely distributed around the world with particularly large reserves in the United States (27.6%), Russia (18.2%) and China (13.3%). Big reserves are also held by India, Australia, South Africa, Kazakhstan and Ukraine (Fig. 4.3).

---

14. With Vladimir Putin's leadership (since 1999), Russian hegemony in the control of natural gas has increased. Gazprom, in fact, is acquiring control of more and more transport infrastructures (Volkov, 2004; Hurst, 2010).

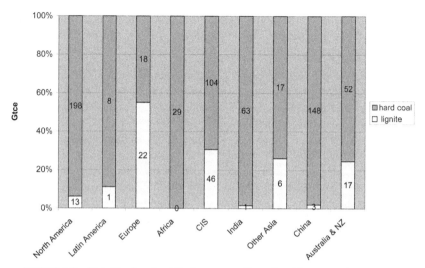

FIGURE 4.3 Global hard coal and lignite reserves. *Reproduced from:* www.bgrcorp.com.

In the coming years, coal demand is expected to remain relatively stable in Russia and Japan and to grow in the United States, India and China. China, in particular, is expected to consume ever-increasing quantities as it struggles to keep up with rising demand for electrical power. As many as 1000 new coal-fired power plants are expected to come on line in China over the next 25 years. India, too, is expected to build many more coal-fired plants in order to satisfy its growing need for electricity. For most of these countries, a pronounced reliance on coal can be explained by its presumed abundance and relatively low cost. Important exporting countries for hard coal are Indonesia, Australia, Russia, the United States, Colombia and South Africa, who together accounted for around 87% of all coal exports in 2012. The top coal importing countries are China, Japan, India, South Korea, Taiwan, Germany, the United Kingdom, Russia, Turkey, Italy and Spain, together accounting for 80% of the coal trade.

Trends in coal use differ from region to region. In Organisation for Economic Cooperation and Development (OECD) countries, coal consumption declined slightly since 2000; in the EU it decreased of about 14%. In contrast, coal demand in developing countries has increased dramatically. Growth in non–OECD countries from 2000 to 2012 amounted to 2.3 Gtce, (+126%). The main driver was China, where coal consumption increased from 1.0 Gtce in 2000 to 2.8 Gtce in 2012. Thus, China has accounted for 83% of the growth in world coal consumption; India accounted for 12%.

On an energy basis, the European Union is the world's fourth largest consumer of coal after China, the United States and India. In the EU, hard coal production has declined from Europe's mature production centres whilst volumes of imported coal have grown significantly. Major coal consuming countries in the region are Germany, Poland, the United Kingdom, the Czech Republic, Italy, France, Greece, Spain, the Netherlands, Bulgaria and Romania. In 2012, Germany was the largest coal importer in the EU, followed by the UK, Italy, Spain, France, the Netherlands and Poland. In 2012, 17% of all coal exports were destined for EU member states. Leading exporters to the EU are Russia (26.1%), Colombia (23.9%), the United States (18.1%), Australia (8.9%), South Africa (8%) and Indonesia (5.2%). Imported hard coal makes a significant contribution to the EU's security of energy supply and offers a competitive fuel which can be easily and safely transported and stocked. Coal offers a much higher level of supply security: the reserves and resources of coal and lignite that are most significant together account for 94% of the EU's remaining potential. Hard coal, both produced and imported, is much less expensive than imported oil or gas and the majority of EU member states enjoy the benefits of coal.

The environmental impacts associated with coal are now fairly understood. Inevitably, coal mining interferes with the environment; however, ecological impacts are increasingly well addressed during mine planning, operation and landscape restoration. The maritime transport of coal is safer than before and it can be easily stocked in large quantities. Emissions from coal use, such as sulphur dioxide, NO_x and dust, can be almost eliminated by commercially available pollution control equipment. In the EU, most coal-fired power plants are now equipped with highly efficient flue gas desulphurisation. For some years, the environmental debate has focused on global climate protection. The strategy to reduce CO_2 emissions from coal use begins with more efficient state of the art power plants, assumes the further development of power plant technology to reach higher efficiencies and leads ultimately to power plants fitted with CO_2 capture and storage. Installations with CO_2 capture should be commercially available by 2020, reducing CO_2 emissions from coal-fired plants by around 90%. Central to the wide use of this technology is an investment friendly legal framework and public support for a CO_2 transport and storage infrastructure.

4.4 EUROPEAN ENERGY GEOPOLITICS: KEY REGIONS

After having introduced the concept of new energy world order, its features and the way the EU and its countries position within the latter, it is worth analysing in detail the main strategic areas for EU energy supply. These strategic areas possess meaningful reserves – oil and/or natural gas – which are crucial in world geopolitics. The strategic areas are the Persian Gulf, the Caspian Sea area, and North-western Africa.[15]

---

15. Of course, other areas may play a major role in world energy geopolitics but, in the logics of this chapter, the discourse focuses on those areas which play a main role according to a European perspective.

4.4.1 The Persian Gulf

The Persian Gulf area includes the coasts of Oman, United Arab Emirates, Saudi Arabia, Qatar, Bahrain, Kuwait, Iraq and Iran. According to British Petroleum, in 2012 the Persian Gulf possessed about 800 billion barrels of oil in proven reserves. To put it differently, this relatively small geographical area possess almost half of the world's oil reserves (BP, 2013). As discussed above, oil reserves are concentrated mainly in Saudi Arabia, Iran, Iraq, Kuwait and the United Arab Emirates; Iran and Qatar also possess huge reserves of natural gas. Some parts of the Persian Gulf are characterised by great instability because of wars, ethnic-religious conflicts and several disputes, such as those concerning Iran's nuclear programme (Merrill, 2007; Barnes and Jaffe, 2006).[16] With the exception of Iran and Iraq, the Countries of the area are grouped in the Cooperation Council for the Arab States of the Gulf (CCASG).

Europe imports a respectable share of oil from the Gulf – approximately 15% of the oil that Europe needs (EC, 2013) – but the main external country with a major political and military influence in the area is the United States and it is not a coincidence that the United States tends to interpret any operation carried out in the area as a potential national or world threat (Klare, 2008a). However, signs of resistance and opposition to American influence over the area occurred cyclically in recent years. It is worth mentioning the position of Abdullah, the Saudi King, in March 2007, who considered illegal the US military occupation of Iraq.[17] The Saudi King affirmed that Arab countries would have to cooperate, all together, to solve the region's problems and to avoid the Americans determining their fate. Moreover, energy reserves in the Persian Gulf are so large that several other countries – Russia, China, India and Japan in particular – are trying to expand their influence here (Barnes and Jaffe, 2006). There is no doubt that growing economic relations with China, Japan, Russia and other countries are in line with the words of the Saudi King and with the attempt to erode the role of the Americans in the Gulf; this process of diversification will intensify, probably, in the future (Goldthau and Witte, 2010).

4.4.2 The Caspian Sea

In the last two decades, the Caspian Sea basin has increased its importance as supplier of oil and natural gas for the world markets: this area, with large untapped oil fields, has been defined as the new 'Great Game' (Gokay, 2006) because of the international competition for the control of its strategic resources.

---

16. In this concern, it is important to mention the historic US–Iran agreement (July 2015) that aims to eliminate the sanctions imposed by the United States to Teheran in exchange for a significant reduction of Iran's nuclear programme.
17. The situation of the area is today very fluid, due to the conflicts and events that characterise it (eg, the recent elections in Iraq, the Syrian war, the role of the Caliphate, etc.) which may lead, in the close future, to the consolidation of an alternative regional order.

At the time of the Soviet Union only two independent states, the Soviet Union and Iran, faced the Caspian Sea basin. Today there are three new ones: Azerbaijan, Kazakhstan and Turkmenistan. Kazakhstan and Azerbaijan possess large reserves of oil, and Turkmenistan is characterised by large natural gas reserves. According to the US Department of Energy (DOE, 2010), this area will have an increase in oil production of about 171% between 2005 and 2030 (from 2.1 to 5.7 million barrels per day). In addition, Turkmenistan is the fourth largest gas exporter in the world (after Iran, Russia and Qatar) and it has about 17.5 trillion m^3 of gas reserves (BP, 2013).

As mentioned above, before the collapse of the Soviet Union, Central Asian and Caucasian states were strictly controlled by Russia: oil and gas were always consumed within the borders of the USSR, and foreign companies were not allowed to operate in this area. Most of the decisions concerning oil platforms, refineries and pipelines were taken by Soviet planners (Gokay, 2001). This scenario changed in the aftermaths of the formation of the independent states of the Caspian Basin in 1991: these countries, generally lacking technical and financial capacities to fully exploit their oil and gas reserves started to cooperate with Western companies in order to break free from Russian control. They allowed foreign companies to extract national oil and gas obtaining, in a few years, foreign direct investments for billions of dollars (Sukhanov, 2005).

The hydrocarbons of the Caspian Sea basin are crucial for many geopolitical actors. Russia traditionally controlled the Caspian Sea basin since the beginning of the 19th century. Currently, United States, Europe and China consider the Caspian Basin an attractive alternative to the Persian Gulf (Gokay, 2006). Most of the existing pipelines in the region have been built by the Soviets during the Cold War years. In recent years, the United States promoted the construction of alternative export routes bypassing the Russian territory; for example, the United States, Azerbaijan and Georgia in 2006 completed the Baku-Tbilisi-Ceyhan oil pipeline, which runs from Baku (Azerbaijan) to Ceyhan (Turkey) avoiding the Bosporus.

The European Commission hopes that the Caspian basin will help to reduce European dependence on Russia (Gokay, 2006). European companies have a substantial presence in some of the largest reserves of the area (for example, Eni and British Gas hold a significant amount of reserves in Karachaganak; Eni, Total, Royal Dutch Shell and Inpex play a key role in the consortium managing Kashagane's fields). European companies are also interested in the construction of pipelines carrying Caspian oil to Europe without passing through the Russia Federation: several European consortia participated in designing corridors, as in the cases of the Nabucco project, or White Stream project, which should transport natural gas from Turkmenistan to Central Europe, running through Ukraine.

It has to be considered that the pipeline strategy is crucial for the Kremlin, too. In past decades, Russia has been fast in realising new corridors in the former USSR countries, like in the case of the Caspian pipeline

(Caspian Pipeline Consortium) linking, since 2001, Tangiz (Kazakhstan) to Novorossiisk (Russia). Russia, by the means of Gazprom and other national companies, signed agreements with main Caspian energy producers and with transit countries in order to manage energy exports directed to Europe (Gokay, 2006).

4.4.3 Africa

Africa is characterised by an abundance of raw materials in a deeply divided continent, with often politically weak countries which are exposed to international exploitation (Watts, 2008). Africa owns some of the largest unexploited oil and gas deposits of the world, as well as extensive reserves of bauxite, cobalt, chromium, copper, platinum, titanium and uranium, mines of gold and diamonds. Because of the world's increasing thirst for energy, Africa is the battleground for a fierce competition between a large number of transnational corporations and countries (Carmody and Owusu, 2007). Some experts argue that African oil will be one of the cornerstones of the energy issue in the coming decades (Ferguson, 2006). International interest for energy resources in Africa is so high that some scholars speak about a new 'scramble for Africa' (Lee, 2006).

With about 10 million barrels per day in 2012, Africa produces about 10% of global oil production (BP, 2013). According to BP, Africa possesses about 126 billion barrels of oil in proven reserves, nearly 10% of the world's total (BP, 2013). Thanks to the discovery of new deposits and the intensive exploitation of existing ones, Africa is the continent with highest growth in oil production, while Africa is also the continent with the lowest level of oil consumption (3.5% of world consumption in 2012). Oil production in Africa is concentrated in the Mediterranean coast (particularly in Algeria and Libya) and, in the last decades, also in the Gulf of Guinea. The oil extracted in this region is considered of excellent quality and the majority of the new fields are located off-shore. Oil extracted off-shore is characterised by lower transport costs. It is also easier to guarantee the security of off-shore sites because they are isolated by political events on the mainland (Ferguson, 2006).

From a geopolitical point of view, Africa is considered by the United States an ideal energy supplier, and African oil represents about 25% of US imports (Carmody and Owusu, 2007). During the last decade also China's dependence on African oil has increased: in 2012, China imported 46 million tonnes of oil from African countries; in 2006, Angola became the main supplier of China's foreign oil, surpassing Saudi Arabia (Carmody and Owusu, 2007). Chinese national companies, like China National Off-shore Oil Corporation (CNOOC), China National Petroleum Corporation (CNPC) and Sinopec, purchased rights for the exploration and exploitation of oil and gas in Angola, Nigeria, Sudan, Gabon, Congo Brazzaville, Equatorial Guinea, Mauritania, Niger, Kenya, Algeria, Libya and Somalia.

European countries are key players in the exploitation of African resources because of geographical proximity, because of old connections dating back to colonial times, and because of the desire to diversify energy suppliers, ie, to weaken the role of Russia as the main energy supplier (Dicken, 2007). The French transnational corporation Total produces oil in seven African countries: Algeria, Angola, Cameroon, Congo-Brazzaville, Gabon, Libya and Nigeria. Total is also the main foreign corporation investing in Congo-Brazzaville and Gabon. Differently, British transnational corporations operate specifically in former colonies. British Petroleum, for example, has invested in Algeria (in alliance with Sonatrach), in Libya (in alliance with the local company National Oil Company) and in Angola (particularly in activities of off-shore extraction). Royal Dutch Shell in 2005 has extracted about 1.1 million barrels per day, right before the closure of many extraction sites as a consequence of riots and disorders in the delta of river Niger. Italian transnational corporation Eni (formerly a state-owned enterprise, currently privatised) has invested in Algeria, Angola, Congo-Brazzaville, Egypt, Libya and Nigeria.

Many lesser-known European transnational corporations have been operating in Africa for decades, in strict connection with local elites and local policy makers. The meaningful profits exploited by foreign corporations, together with the lack of positive spillovers for African economies, have been at the centre of a number of critical analyses (Dicken, 2007). European corporations are willing to maintain their hegemonic positions in the continent in the future, but during the last decade their role in the extractive industry has been reduced because of the growing role of giant American energy corporations and because of the investments from China and India. Overall, the global competition for the control of African energy resources underlines a number of global problems concerning global energy governance and, in general, the uneven, ongoing processes of economic globalisation (Ferguson, 2006; Dicken, 2007).

4.5 CONCLUSIONS AND FUTURE PERSPECTIVES

The aim of this chapter has been to build a geographical analysis of political relations, paths, risks and possibilities related to European energy security. In order to explore this topic, the geographies of oil, natural gas and coal supplies have been analysed, and current bottlenecks, potential dangers and possible lines of development have been discussed.

The main conclusions deriving from the above sections may be summarised in specific geographical representations emphasising the main directions that Europe needs to follow for future energy security. The production of representations and scenarios is indeed a typical task of disciplines as geography and spatial planning. Representations are devices for interpreting reality, for communicating messages, for thinking about alternatives and for building politics. Political choices are not rarely based on symbolic representations of space, such

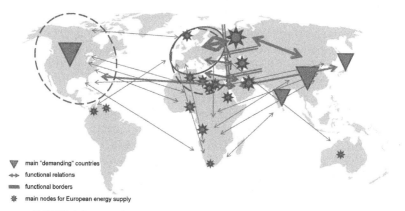

main "demanding" countries
functional relations
functional borders
main nodes for European energy supply

FIGURE 4.4 A spatial scenario European energy security. *Authors' own elaboration.*

as spatial metaphors or visual metaphors (Barnes and Duncan, 1992).[18] The kind of representations proposed in this conclusive section are explicitly suggestive and communicative, by taking advantage of the use of chorematic diagrams, ie, dynamic symbols introduced in maps in order to summarise information and to show relevant (supposed or actual) spatial dynamics (see the classic work of Reynaud, 1981; see also Reimer, 2010; Paklone, 2011).[19]

The map proposed in Fig. 4.4 is a first attempt to build a geographical representation of Europe's view of macroregional geopolitics of energy security. In the picture, the stars represent the main nodes for European energy supply, while the triangles are the main 'demanding' countries. As discussed in the previous sections, the main strategic areas are Russia, the Caspian Sea area and the Persian Gulf area. While the Caspian and the Persian areas may be conceptualised as spaces characterised by a certain internal coherence (the Caspian countries willing to emancipate from Russian hegemony; the OPEC countries may be considered as a sociological 'collective actor'), as well as the evident case of Russia, this is definitely not the case for Africa. Africa is in fact characterised by a high degree of internal fragmentation: a number of countries have developed individual relations with Europe or with single European countries, as well as with the United States and/or China. Africa is indeed a contested space, at the centre of strategic fluxes and investments from all over the world.

---

18. It has to be mentioned that every spatial representation is embedded in subjective choices, concerning, for example, the use of conventional symbols, colours and scales (Harley, 1989; Starling, 1998). In this sense, geographical representations have to be evaluated not just in terms of accuracy, but above all in terms of usefulness: are geographical representations useful for the circulation of knowledge and the building of innovative ideas? Are geographical representations useful for political consensus? How communicative are they (Taylor et al., 1995; Vanolo, 2010)?

19. Probably the best known examples are the maps and scenarios proposed by French institution DATAR (see for example Datar, 2000) or, going back in time, the famous 'blue banana' proposed by French geographer Brunet (1989). Today, visual scenarios based on qualitative hypothesis are widely used in planning activities all over Europe (Dühr, 2003, 2007; see also CRPM, 2002).

From a quantitative point of view, energy flows are currently not as important as those involving Russia, the Caspian Sea area and the Persian Gulf area, but flows are important because geopolitical scenarios are still open, and it is not yet so clear if and how Europe, the United States, China or other rising powers (such as India) will secure their energy supplies from Africa.

In the African space, only the northern part of the continent is strongly connected to Europe in terms of safe energy flows (larger European chorem, with dotted line). The Mediterranean area has long-time energy relations with Southern Europe, and these relations are considered relatively stable. In a similar way, gas and oil provisions from the North Sea are highly safe (and this is the reason because the North Sea star-symbol has been put inside the European chorem), but rather marginal from a quantitative point of view.

As is well known, the Persian Gulf area (and similarly, on a smaller scale, also the Caspian Sea area) is a global energy supplier, and for this reason meaningful connections link the Persian Gulf with major global energy importers, as the United States, Europe, China and India.

As fully discussed in the previous section, the key node for European energy security is Russia. Russia is the biggest 'star' in the qualitative representation of Fig. 4.4, because Russia is de facto the main European energy supplier. At the same time, 'border effects', that means (in this case) conflictual connections, characterise the relations with Russia, as well as between Russia and the countries of the Caspian area (double-red lines, in the figure). Russia plays indeed a quasi-monopolistic role in relation to Europe and, at the same time, has a sort of 'imperialist' attitude towards its southern neighbours. A key element for the energy security in Europe is to diversify as much of the energy supply as possible, while of course building stable and friendly relations with Russia.

Fig. 4.5 is a representational exercise dealing with a 'maximum diversification' scenario. It graphically represents the optimistic idea that Europe, in the future, will diversify as much as possible the geography of its energy suppliers. The figure proposes three different types of arrows. The thicker arrows are 'first level suppliers'. These are the geographical relations that will be pivotal for European energy supply. In this hypothetical scenario, Russia will still be a major supplier, but a number of other areas will share a similar role. In particular, the Caspian area and the Persian basin area will export oil and gas side by side with Russia (but independently from the latter). Of course, such an option will be possible if the ongoing 'pipeline war', discussed in previous sections, will end with a major role of European companies in the control of transport corridors. Finally, Africa will be a major oil exporting area for Europe.

Secondary suppliers are geographical spaces that will play a minor role in oil provision. Of course, a minor role may still be important for diversification. The only area represented as a secondary supplier is Central America. Currently, Venezuela is a European oil supplier; it is not likely that its role will increase in the future, but it is a player in global energy geopolitics. Finally, the potential development axis refers to connections that are currently not relevant, but that may become crucial in the future. To put it differently, potential development

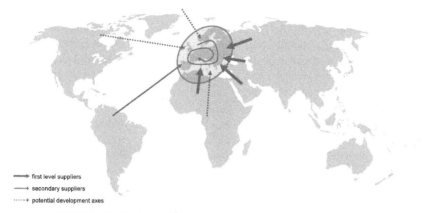

first level suppliers
secondary suppliers
potential development axes

FIGURE 4.5 Maximum diversification scenario. *Authors' own elaboration.*

axis may represent European challenges in the future. For example, the connection with Canada is represented. Canada is right now a minor oil supplier for Europe, but its role may increase in the future if new technologies will help the exploitation of local giant reserves of bituminous sands. The Western Arctic reserves will probably become a key geopolitical area because of the major oil reserves located in the area; it is estimated that about one-fourth of the world's unexplored oil reserves are there. Finally, a key potential development axis connects Europe with sub-Saharan territories in Africa; a key area is, of course, the Guinea gulf, that is right now a contested space for the control over oil extraction.

Of course, Europe's energy self-reliance will mean increasing internal energy production, and energy efficiency is pivotal in this scenario. The more the EU becomes self-reliant, the more it will be resilient and resistant to energy crises and fluctuations in energy markets, reducing at the same time the need for the diversification of external suppliers. Currently, the self-reliance scenario seems to be impossible, but internal production and energy efficiency have to be considered as virtuous processes: if self-reliance is (actually) impossible, it doesn't mean that it is not necessary to try to walk that path.

It is possible to extrapolate some additional considerations. First and foremost, it is crucial to reduce energy dependency, which means increasing internal energy production and energy efficiency (see Chapter 2). Technology is the most obvious way to achieve this goal, and in this sense the role of research and development (R&D) is confirmed as a cornerstone element. The EU is certainly pushing R&D in many ways – consider, for example, the recent emphasis on smart city programmes – and this chapter emphasises how a decrease in the need for oil and natural gas may free Europe from a number of complex and evolving geopolitical struggles. It has to be mentioned that the controversial possibility of increasing nuclear energy production in Europe may be considered as an alternative path to foster self-reliance in the energy field as clean coal technologies implementation will be a self-reliance alternative too.

Second, reducing the Russian hegemonic position for European energy supply is a hot topic. A geographical diversification of energy imports is needed in order to make European energy supply safer, more reliable and probably also more competitive and resilient.

Third, it should be useful to adopt, for the EU, a constructive approach through a policy framework around elements of the enlightened self-interest of those governments displaying features of State Capitalism, such as China. In the case of some countries which are part of a transition from the Liberal Capitalist to the State Capitalist regime, and only so far display limited State Capitalist tendencies, it may even be feasible to arrest those tendencies by pro-actively pursuing mutual interests on a country-to-country basis. This would imply, for example, a higher than hitherto level of joint investment, accelerated adoption of common technical, legal, commercial and market standards, and a concentration on truly open trade based on mutual understanding and advantage, rather than on a culture of complaint and counter-complaint. The pursuit of mutual self-interest should result in a more secure and a more dependable access to oil, gas and minerals.

Finally, it has to be mentioned that this chapter focused on the external geopolitical dimension of energy security, particularly by focusing on the problem of energy provision, and therefore the report has not considered a number of topics that have been already analysed in Chapter 2 as the need for developing an internal integrated energy market and the relations between energy production and climate change. These perspectives are complementary but important takes on the European energy question.

REFERENCES

Aalto, P. (Ed.), 2008. The EU-Russian Energy Dialogue: Europe's Future Energy Security. Ashgate, Aldershot.

Acaravci, A., Ozturk, I., 2010. On the relationship between energy consumption, CO_2 emissions and economic growth in Europe. Energy 35 (12), 5412–5420.

Acuto, M., 2010. High-rise Dubai urban entrepreneurialism and the technology of symbolic power. Cities 27 (4), 272–284.

Apergis, N., Payne, J.E., 2010. The emissions, energy consumption, and growth nexus: evidence from the commonwealth of independent states. Energy Policy 38 (1), 650–655.

Arouri, M.E.H., Youssef, A.B., Mehenni, H., Rault, C., 2012. Energy consumption, economic growth and CO_2 emissions in Middle East and North African countries. Energy Policy 45, 342–349.

Badalyan, L., 2011. Interlinked energy supply and security challenges in the south Caucasus. Caucasus Analytical Digest 33.

Barnes, T.J., Jaffe, A.M., 2006. The Persian Gulf and the geopolitics of oil. Survival 48, 143–162.

Barnes, T.J., Duncan, J.S., 1992. Introduction. Writing worlds. In: Barnes, T.J., Duncan, J.S. (Eds.), Writing Worlds. Discourse, Text & Metaphor in the Representation of Landscape. Routledge, London, pp. 1–17.

Behr, P., 2010. Energy nationalism: do petrostates threaten global energy security? CG Global Researcher 1 (7), 151–180.

Bentley, R.W., 2002. Global oil & gas depletion: an overview. Energy Policy 30 (3), 189–205.

Bilgin, M., 2009. Geopolitics of European natural gas demand: supplies from Russia, Caspian and the middle east. Energy Policy 37 (11), 4482–4492.

Bosse, G., 2011. The EU's geopolitical vision of a European energy space: when 'Gulliver' meets 'white elephants' and Verdi's Babylonian king. Geopolitics 16 (3), 512–535.

(BP) British Petroleum, 2013. Statistical Review of World Energy 2013. http://www.bp.com/en/global/corporate/about-bp/energy-economics/statistical-review-of-world-energy-2013.html.

Bradshaw, M.J., 2009. The geopolitics of global energy security. Geography Compass 3 (5), 1920–1937.

Brunet, R., 1989. Les Villes Européennes, Datar-reclus. La Documentation Française, Paris.

Carmody, P.R., Owusu, F.Y., 2007. Competing hegemons? Chinese versus American geo-economic strategies in Africa. Political Geography 26 (5), 504–524.

Champion, M., 2006. Russian energy grip splits EU. Wall Street Journal. November 13th, Available http://online.wsj.com/news/articles/SB116338183831721008.

Cohen, S.B., 1991. Global geopolitical change in the post-Cold War era. Annals of the Association of American Geographers 81 (4), 551–580.

Conférence des Régions Périphériques Maritimes d'Europe (CRPM) (Ed.), 2002. Study on the Construction of a Polycentric and Balanced Development Model for the European Territory. CRPM, Rennes.

Cornelius, P., Story, J., 2007. China and global energy markets. Orbis 51 (1), 5–20.

Correlje, A., Van der Linde, C., 2006. Energy supply security and geopolitics: a European perspective. Energy Policy 34 (5), 532–543.

Curtis, F., 2009. Peak globalization: climate change, oil depletion and global trade. Ecological Economics 69 (2), 427–434.

Dabashi, H., 2012. The Arab Spring: The End of Postcolonialism. Zed Books, London.

Datar, 2000. Aménager la France de 2020. Mettre les territoires en mouvement. La Documentation française, Paris.

Department of Energy United States (DOE), 2010. International Energy Outlook 2010. http://www.energy.gov/.

Dicken, P., 2007. Global Shift. Mapping the Changing Contours of the World Economy. Sage, London.

Dühr, S., 2003. Illustrating spatial policies in Europe. European Planning Studies 11 (8), 929–948.

Dühr, S., 2007. The Visual Language of Spatial Planning. Exploring Cartographic Representations for Spatial Planning in Europe. Routledge, London.

(EC) European Commission, 2006. A European Strategy for Sustainable, Competitive and Secure Energy. Green Paper, EC.

(EC) European Commission, 2013. EU Energy Statistics Pocketbook 2013. http://ec.europa.eu/energy/observatory/statistics/statistics_en.htm.

(EIA) Energy Information Administration, 2010. Annual Energy Outlook 2009. EIA.

(EIA) Energy Information Administration, 2013. Annual Energy Outlook 2012. EIA.

Favennec, J.P., 2011. The Geopolitics of Energy. Technip Editions, Paris.

Ferguson, J., 2006. Global Shadows. Africa in the Neoliberal World Order, Durham and London. Duke University Press.

Gerboni, R., Grosso, D., Schranz, L., 2014. REACCESS EU Project Outcomes Elaboration. Politecnico of Turin.

Gokay, B. (Ed.), 2001. The Politics of Caspian Oil. Macmillan, London.

Gokay, B., 18 May 2006. The Power Shift to the East: The 'American Century' Is Ending. Pravda.

Goldthau, A., Witte, J.M. (Eds.), 2010. Global Energy Governance: The New Rules of the Game. Global Public Policy Institute, Berlin.

Goodstein, D.L., 2004. Out of Gas. W.W. Norton, New York.

Hall, C.A., Day, J.W., 2009. Revisiting the limits to growth after peak oil. American Scientist 97 (3), 230–237.

Harley, B., 1989. Deconstructing the map. Cartographica 26 (2), 1–20.

Helm, D., 2002. Energy policy: security of supply, sustainability and competition. Energy Policy 30 (3), 173–184.

Hisane, M., 2006. Japan's New Energy Strategy. Asia Times. January 13th, Available http://www. atimes.com/atimes/Japan/HA13Dh01.html.

Hurst, C., 2010. The militarization of Gazprom. Military Review 90 September–October.

(IEA) International Energy Agency, 2004. Energy to 2050: Scenarios for a Sustainable Future. Pa.

(IEA) International Energy Agency, 2013. World Energy Outlook 2013 Annual Report. Paris.

(IPCC) Intergovernmental Panel on Climate Change, 2008. Energy Technology Perspectives. Scenarios and Strategies to 2050. OECD/IEA, Paris.

(IPCC) Intergovernmental Panel on Climate Change, 2000. Special Report on Emissions Scenarios (SRES). Summary for Policymakers, a Special Report of IPCC Working Group III.

(IPCC) Intergovernmental Panel on Climate Change, 2007. Fourth Assessment Report: Climate Change 2007. Synthesis report, Geneva.

Jojarth, C., 2008. The End of Easy Oil: Estimating Average Production Costs for Oil Fields Around the World, Center on Democracy, Development, and the Rule of Law-Stanford. Program on Energy and Sustainable Development Working Paper, 72.

Kerr, R.A., 2011. Peak oil production may already be here. Science 331 (6024), 1510–1511. http:// dx.doi.org/10.1126/science.331.6024.1510.

Klare, M., 2001. Resource Wars: The New Landscape of Global Conflict, New York. Metropolitan Books.

Klare, M., 2004. Blood and Oil: The Dangers and Consequences of America's Growing Dependency on Imported Petroleum, New York. Metropolitan Books.

Klare, M., 2008a. Rising Powers, Shrinking Planet. Oneworld, Oxford.

Klare, M., 2008b. The New Geopolitics of Energy. The Nation. May 19th, Available http://www. jmhinternational.com/news/news/selectednews/files/2008/05/20080501_Nation_%20The-NewGeopoliticsOfEnergy.pdf.

Laconte, P., 2011. Climate Change, Energy Storage, Biodiversity Loss. An Overview of Global, European and Local Policies and Practices. The Club of Rome – European Support Centre and EU Chapter.

Laurmann, J.A., 1992. World energy prices, geopolitics and global warming. International Journal of Hydrogen Energy 17 (7), 553–554.

Lean, H.H., Smyth, R., 2010. CO_2 emissions, electricity consumption and output in ASEAN. Applied Energy 87 (6), 1858–1864.

Lee, M.C., 2006. The 21st century scramble for Africa. Journal of Contemporary African Studies 24 (3), 303–330.

Li, Z.D., 2003. An econometric study on China's economy, energy and environment to the year 2030. Energy Policy 31 (11), 1137–1150.

Merrill, K.R., 2007. The Oil Crisis of 1973–1974: A Brief History with Documents. Bedford/St. Martin's.

Paklone, I., 2011. Conceptualization of visual representation in urban planning. Limes: Cultural Regionalistics 2, 150–161.

Pao, H.T., Tsai, Ch. M., 2010. CO_2 emissions, energy consumption and economic growth in BRIC countries. Energy Policy 38 (12), 7850–7860.

Peters, S., 2004. Coercive western energy security strategies: 'resource wars' as a new threat to global security. Geopolitics 9 (1), 187–212.

Reimer, A.W., 2010. Understanding chorematic diagrams: towards a taxonomy. The Cartographic Journal 47 (4), 330–350.

Reynaud, A., 1981. Société, espace et justice. PUF, Paris.

Roberts, P., 2004. The End of Oil: On the Edge of a Perilous New World. Houghton Mifflin, Boston.

Selley, N., 2013. The New Economy of Oil: Impacts on Business, Geopolitics and Society. Routledge, London.

Soytas, U., Sari, R., 2009. Energy consumption, economic growth, and carbon emissions: challenges faced by an EU candidate member. Ecological Economics 68 (6), 1667–1675.

Starling, R., 1998. Rethinking the power of maps: some reflections on paper landscapes. Ecumene 5 (1), 105–108.

Stern, J.P., 2005. The Future of Russian Gas and Gazprom. Oxford University Press, Oxford.

Stern Review, 2006. Review on the Economics of Climate Change. UK Office of Climate Change.

Sukhanov, A., 10 February 2005. Caspian Oil Exports Heading East. Asia Times.

Taylor, P.J., Watts, M.J., Johnston, R.J., 1995. Remapping the world: what sort of map? what sort of world? In: Johnston, R.J., Taylor, P.J., Watts, M.J. (Eds.), Geographies of Global Change. Remapping the World in the Late Twentieth Century. Blackwell, Oxford, pp. 377–385.

Toft, P., Duero, A., Bieliauskas, A., 2010. Terrorist targeting and energy security. Energy Policy 38 (8), 4411–4421.

Umbach, F., 2010. Global energy security and the implications for the EU. Energy Policy 38, 1229–1240.

(UNDP) United Nations Development Programme, (UNDESA) United Nations Department of Economic and Social Affairs, (WEC) World Energy Council, 2001. World Energy Assessment: Energy and the Challenge of Sustainability. New York.

(UNEP), United Nations Environmental Program, 2006. GEO: Global Environment Outlook 3. Past, Present and Future Perspectives. Washington.

Vanolo, A., 2010. The border between core and periphery: geographical representations of the world system. Tijdschrift voor Economische en Sociale Geografie 101 (1), 26–36.

Victor, D.G., Jaffe, A.M., Hayes, M.H. (Eds.), 2006. Natural Gas and Geopolitics: From 1970 to 2040. Cambridge University Press, Cambridge.

Vivoda, V., 2010. Evaluating energy security in the Asia-Pacific region: a novel methodological approach. Energy Policy 38 (9), 5258–5263.

Volkov, V., 2004. Hostile enterprise takeovers: Russia's economy in 1998–2002. Review of Central and East European Law 29 (4), 527–548.

Watts, M.J. (Ed.), 2008. Curse of the Black Gold: 50 Years of Oil in the Niger Delta. Powerhouse Books, Brooklyn, NY.

Winzer, C., 2012. Conceptualizing energy security. Energy Policy 46 (C), 36–48.

Youngs, R., 2009. Energy Security: Europe's New Foreign Policy Challenge. Routledge, London.

Chapter 5

Reshaping Equilibria: Renewable Energy Mega-Projects and Energy Security

M. Gruenig[1,2], B. O'Donnell[2]
[1]Ecologic Institute, Berlin, Germany; [2]Ecologic Institute US, Washington, DC, United States

5.1 INTRODUCTION: ASSESSING THE CONTRIBUTION OF LARGE-SCALE RENEWABLE ENERGY PROJECTS TO LOW-CARBON ENERGY SECURITY

The energy sector's mantra for the past century has been to build capacity. With the coupling of economic growth and energy use, whether for large-scale production, trans-oceanic and intercontinental supply chains or increased final consumption, national and regional economies have depended on increasing energy demand. In order to provide security for such an energy system, low-carbon energy policies have continued to function within this paradigm, concentrating foremost on building capacity from renewable sources, usually distributed within existing transmission schemes with the objective being to match and eventually replace carbon-based energy supplies. The gap between renewable energy production, based on replacing previous demand, and the growth in demand, as necessitated by the continued adherence to this paradigm, has produced an opportunity for carbon-based energy sources to maintain a position in the energy security agendas.

A seemingly forward-thinking approach is to create mega-projects with renewable energy sources (RES) that would potentially produce more energy than the world could ever use. That has long been the argument in favour of solar energy, for instance: there is no quantifiably greater source of energy than the sun. But abundance of a source does not guarantee efficiency of use, as we can see from the current debates surrounding stranded assets in the carbon market. Additionally, the transmission and distribution networks in Europe are still lagging behind planned integration strategies (this will be further discussed in Chapter 6). Moreover, and perhaps most relevant to this chapter, renewable sources in great quantities are seldom found in or near centres of consumption,

Low-carbon Energy Security from a European Perspective. http://dx.doi.org/10.1016/B978-0-12-802970-1.00005-X
Copyright © 2016 Elsevier Ltd. All rights reserved.
109

increasing waste through transmission and maintaining a system which leaves networks and consumers vulnerable to a variety of disruption potentials.

Renewable energy mega-projects, such as those that will be discussed in this chapter, are likely to play a role in long-term low-carbon energy security in Europe. However, questions must be asked to ensure we are not simply exchanging inputs to an existing insecure system. What aspects of an energy system are most vulnerable? How can those vulnerabilities be minimised? And how can energy policy shift paradigms, trading capacity for efficiency, making reliability and security synonymous and producing an energy secure future for Europe?

This chapter will look at a range of large-scale as well as distributed approaches to renewable energy and will assess their respective potential contributions towards energy security, specifically low-carbon energy-security. While energy security encompasses security for all forms of energy, this chapter will focus exclusively on electricity security. This approach is broadly supported by the expected increase in electricity in the final energy use due to electrification of transport, insulation of buildings and resulting in reduced heating demand, increased use of heat pumps for residual heating demand as well as the use of low-cost excess renewable production for industrial application such as steel or aluminium mills.

In the context of this chapter, we define large-scale projects as energy generation projects with significantly over 100 MW generating capacity.

Bearing in mind the very different nature of the projects and proposals under examination in this chapter, it is important to apply a common reference frame.

Building on the Milesecure-2050 project (Crivello et al., 2013), this chapter assesses the contribution of energy visions and projects based on the following framework:

i. **Stability**

The capacity of the highly connected energy system to maintain its operation within acceptable technical constraints, facing sudden disruptions of critical system components affecting transmission and distribution on the seconds to minutes scale.

ii. **Flexibility**

The ability of a system to cope with the short-term uncertainty of the energy system variables by balancing any deviations between the planned or forecasted supply and demand on the one side and the actual situation on the other side over the minutes-to-hours horizon.

iii. **Resiliency**

The energy system can source alternative modes of production or consumption in response to sudden and transient shocks, ie, high impact–low probability events such as the interruption of a major supply source on the hours-to-weeks time horizon.

iv. **Robustness**

The reasonable expectation that the system as a whole is able to meet all demand at all times under all anticipated conditions, taking into account market conditions and the regulatory regime, eg, market failures or faulty market design, over a weeks-to-years horizon.

v. **Adequacy**

Actors in the energy market are allowed to choose from primary energy sources at cost-oriented prices, without being hindered in their choice by economic or (geo)-political constraints or enduring pressures on energy sources and infrastructures over a years-to-decades horizon.

5.2 TAPPING THE DESERT FOR SUSTAINABLE POWER: MOROCCO AND NORTH AFRICA

It sounds like a reasonable idea: the sun pours potential energy onto the planet's deserts all day. Why not harness that potential to fuel the world's cities, producing economic opportunity for the local areas, creating a foundation for international cooperation and integration and helping lead the global low-carbon energy transition? The idea also sounds eerily familiar: large pools of energy (oil) have been just lying beneath the Earth's surface for millennia. Why not use nature's abundance to fuel economic growth, create jobs and unleash the potential of the future?

Clearly, the analogy is not perfect. Replacing carbon energy with RES, including from the sun, is essential to the health and vitality of the planet. However, the analogy does show that concepts for energy production and distribution continue to be caught in the same general paradigm: go big and go far. This paradigm, for constitutional reasons, has led Europe to its current situation, one that is unsustainable and insecure.

Nonetheless, large-scale solar energy projects are not inherently counterproductive to developing a low-carbon energy security, at least outside of Europe. Inside the EU, it is difficult to see where and how large-scale solar could be implemented to produce enough energy to make the financial burdens palatable within a reasonable time horizon. In the Middle East and North Africa (MENA), however, the sun's abundance would seem to be the perfect solution.

And it is. The question, though, is for whom and how. In this chapter, two MENA projects are analysed to assess whether they will increase the potential for low-carbon energy security in Europe. The answer is yes, but not in the way one might expect.

5.2.1 DESERTEC and Noor Project Backgrounds

DESERTEC is the name for a concept to install large-scale concentrated solar-thermal power (CSP) in the North Sahara and export a significant share of the resulting electricity towards Europe via high voltage direct current sea cables (DESERTEC Foundation, 2010).

The idea to harness solar power from African deserts first emerged around 1986 by Gerhard Knies in the aftermath of the Chernobyl nuclear disaster, but its origins can be traced even further back to around 1913 (Hickman, 2011b). Around the same time as Knies' suggestion, the concept of a 'Mediterranean Transmission Line Ring' was discussed by Egyptian minister, Maher Abaza in 1987 (El Nokrashy, 2005).

Knies then initiated the Trans-Mediterranean Renewable Energy Cooperation (TREC) network of researchers with the aim of further developing the concept (Kabariti et al., 2003) which would eventually, with the German section of the Club of Rome, become the DESERTEC Foundation in 2009 (DESERTEC Foundation, n.d.).

In order to accelerate the implementation, a second entity was created in 2009 as well, the DESERTEC industrial initiative (Dii), which grew to 21 shareholders and 35 associated partner companies from 16 countries in Europe as well as MENA (Zickfeld et al., 2012).

The political changes in the region associated with the Arab Spring led to a full stop on both the overall DESERTEC vision and Dii, with significant partners leaving the initiative. However, both the DESERTEC Foundation and Dii are still operating and so are local partial implementations of solar-thermal power in the Sahara such as at Noor in Morocco.

The DESERTEC plan's key aspects are as follows (Zickfeld et al., 2012):

1. integrated EU–MENA power system, including market integration;
2. Europe importing up to 20% of its electricity demand from MENA;
3. 100 GW combined wind and CSP by 2050, supported by balancing gas power plants, in MENA; and
4. total investment needs upward of €400 billion.

The primary arguments for implementation in North Africa are:

- high solar insolation resulting in high photovoltaic (PV) and solar-thermal potentials;
- low population density resulting in low local energy consumption; and
- low land-use competition.

Criticism of the DESERTEC initiative came early and with force especially, but not solely, from civil society in North Africa (Friedman, 2011). The main points of contention centred on the neocolonialist presumptions of the project's European supporters, the lack of local participation and the disparity in benefits for local communities. Tunisian trade unionist Masour Cherni was among the critics of DESERTEC at the 2013 World Social Forum held in Tunis, questioning local benefits and water consumption, among other aspects (Kwasnieswski, 2013).

In August 2013, Dii came to terms with its neocolonial *haut goût* and abandoned the idea of promoting renewable energy exports to Europe and rather instead decided to focus on developing local markets for renewable energy (Euractiv, 2013).

Nevertheless, critics are still abundant and suspicions persist, particularly regarding the dominant role of international corporations, the low technology

FIGURE 5.1 Ouarzazate solar complex. *Mapbox, OpenStreetMap for the administrative boundaries; Cartography: Ecologic Institute.*

transfer and the lack of local community involvement (Hamouchene, 2015). In a wider sense, the neocolonial perception was further deepened by tensions surrounding the parallel creation of the Union for the Mediterranean (Hinnebusch, 2012), the successor to the Barcelona Process, launched in 2008 (French Government, 2008).

However, the prospect of renewable energy generation in MENA is not dependent on the success of DESERTEC. In startling contrast, Morocco's Noor Project has successfully begun harnessing the ubiquitous desert sun to produce energy, though not with the intent to transmit it to Europe. Yet the Noor Project may well do more to facilitate European low-carbon energy security.

The first pilot implementation plant is the Noor CSP near Ouarzazate, Morocco (Hickman, 2011a; Fig. 5.1). Planning of phase I began in 2013, with USD 240 million financing by the African Development Bank and implementing partner Moroccan Agency for Solar Energy (African Bank for Development, 2013). In phase I, the plant has a generating capacity of 160 MW, later to be boosted to 500 MW in phase II (African Bank for Development, 2014). Second phase financing by the World Bank will be USD 400 million. Public–private partnerships are expected to raise the remaining USD 2.2 billion with expected finalisation in 2020 (Mobarek, 2015).

There is no plan to connect the plant directly via new cables with the European power grid. However, the current connection between Spain and Morocco could be improved and could allow a fraction of the electricity produced at Noor to enter the European grid. Still, with the Spanish peninsula poorly interconnected with the remainder of the European electricity market, transmission beyond the Pyrenees would require significant additional infrastructure investments.

A door-opener could have been the Mediterranean Electricity Ring or MedRing project which studied the potential for a Mediterranean energy grid as proposed by the European Commission in 2008 (MED-EMIP, 2010). However, the idea did not survive the turmoil of the Arab Spring and is not part of the Energy Union's projects of common interest.

Noor II is planned to be a 200 MW parabolic through station, while Noor III is designed to be a 150 MW solar tower. Each station will be equipped with thermal storage: Noor II with 2800 MWh corresponding to 5 h of production at full capacity; Noor III with 2730 MWh corresponding to 7 h of production. While Noor I runs on wet cooling, both Noor II and Noor III are designed for dry cooling, reducing total water consumption of the plant (African Bank for Development, 2014). Regardless, water consumption for Noor will be 2.5 to 3 million m³. Moreover, the installation will require between 15,000 t and 17,000 t eutectic salts for thermal storage, as well as synthetic oil for the parabolic through heat transport. Leakages and spill can result in soil pollution; impacts on groundwater have not been assessed. In order to keep the salt liquid, 19 t gasoil will be required per day for the full 500 MW installation. In addition, water for cooling will be brought in initially via trucks, resulting in additional fossil fuel consumption.

The assessment found that the parabolic throughs have a significantly higher land use impact compared to the solar tower: $434,000 \, m^2$ versus $35,000 \, m^2$. Moreover, pollution risk is higher with parabolic throughs due to the heat convecting synthetic oil which also constitutes a significant fire risk (African Bank for Development, 2014). While both designs impact birds and wildlife, the solar tower's high heat flux and glaring is considered far more dangerous for birds than the CSP parabolic through design.

The 2500 ha affected by the project formerly belonged to the Ait Oukrour Toundout ethnic group. The site was used for pastoral herding and is no longer accessible to herders. However, alternative herding areas exist in the vicinity. Other potential social impacts have been further investigated in a study (Terrapon-Pfaff et al., 2015), finding that the local population responded primarily positively to the planned development. While the authors were very aware of the project's neocolonial aspects and the potential risk in implementation, they attribute the high levels of local acceptance to the work of the local Moroccan Agency for Solar Energy.

5.2.2 Middle East and North Africa Region and Moroccan Energy Policies and Support Mechanisms

Energy policy developments in the MENA region are diverse and complex as the region is often split into net oil-importers, such as Morocco, and net oil-exporters such as Saudi Arabia. However, low-carbon energy development throughout the region has gained surprising momentum in the last decade. Focusing attention on the Mediterranean and North Africa region in general and Morocco in particular, the following advancements show a definite progressive trajectory (IRENA, 2013):

- Nonhydro renewable electricity in the region more than doubled from 1.2 TWh to nearly 3 TWh. Growth was particularly strong in solar PV with average annual increases of 112% compared to fossil fuelled electricity growing at a mere 6% on average.
- By 2012, total installed renewable energy capacity amounted to over 19 GW, of which 16.5 GW in net oil-exporting countries. The overwhelming majority of this capacity is still in hydropower, 17.7 GW. Concentrated solar power was just 182 MW in 2012.
- Installed wind power also grew substantially from 260 MW in 2005 to over 1 GW in 2012.
- In April 2013, the renewable energy project pipelines in the region added up to 7.5 GW in additional capacity, of which 4.2 GW were in net oil-importing countries, including 1.7 GW in Morocco.

These growth trends indicate where the MENA region is moving, partly driven by the market, partly by energy policies. By mid 2013, all 21 MENA countries had introduced a renewable energy target, compared to only 5 in 2007, and 18 countries had renewable energy support policies, the vast majority being through public financing (12 countries), followed by tax incentives (7) and also net metering (7).

Renewable energy as a share of electricity supply targets vary substantially across the region from 42% of installed capacity by 2020 in Morocco to 2% of generation in Qatar.

Looking at total renewable energy capacity targets, both net oil-exporters and importers set high benchmarks:

- Saudi Arabia aims for 54 GW by 2032 (now postponed to 2040, see below)
- Algeria aims for 12 GW by 2030
- Egypt aims for 10.7 GW by 2027
- Iraq aims for 7.7 GW by 2016
- Kuwait for 7.7 GW by 2030
- Morocco aims for 6 GW by 2020, the highest target of net oil-importing countries

If these targets are achieved, total projected renewable energy capacity by 2030 would add up to 107 GW. However, these numbers are not legally binding, and, despite the achievements already made, policy changes can happen overnight. A good example is Saudi Arabia which impressed many sceptics with its ambitious targets published in 2012, only to postpone those targets in early 2015 by a full 8 years (Reuters, 2015). Nevertheless, investments in renewable energy in the region grew from USD 474 million in 2009 to USD 2.9 billion in 2012.

Morocco is considered a leader in creating a local renewable energy sector, prioritising the manufacturing of components for solar PV, CSP and solar water heating. With increased investments in education and workforce training, Morocco is building the necessary pillars for the often-promised future-focused green economy. Finally, local content is a criterion in renewable energy tenders, resulting in high shares of local added value, including at Noor.

5.2.3 Energy Security Implications

It is important to recognise the value of large-scale solar energy initiatives in the overall effort to mitigate climate change and create a global low-carbon energy transition. The efforts in MENA, whether DESERTEC or Noor, are rooted in the similar principle of seeking areas of high concentration of an RES, building the facilities to collect it and produce from it consumable energy and transmitting it efficiently to consumption centres where consumers are able to pay the market price, which makes the investment and continued operation competitive. Particularly in MENA, where carbon-intense energy sources have long controlled the energy market both for net-export and net-import countries, the barriers would seem to be especially cumbersome. However, as discussed in the previous section, there is considerable momentum for low-carbon energy developments in the region, and, as the Noor project shows, these development initiatives can be used for broader social and economic benefits.

That being said, the focus of this book is low-carbon energy security in Europe. Therefore, in assessing the energy security implications of the proposed DESERTEC and Noor projects specifically and large-scale solar installations generally, the objective is to use the available lessons and information to consider impacts to Europe's energy system. Since energy systems comprise more than collection, production and distribution strategies, overall market considerations should be taken into account, leading to broader questions as to how the global transition to low-carbon energy security might possibly improve or even impede Europe's efforts.

As a final note, DESERTEC remains hypothetical, making any assessment as to the proposal's productivity presumptive. Lessons from the policy and planning phases, however, do provide opportunities for evaluation and can serve to inform development of future large-scale solar energy projects.

i. Stability

Sourcing and distribution diversification is a chief component of system stability. By this measure, DESERTEC's infrastructure interconnectivity would strongly benefit the European grid. Architecture development between North Africa and Europe would serve as a foundation for further system upgrades, potentiating long-term viability and versatility. The expansion of the grid in Europe would also present opportunities for transmission enhancement of existing regimes, particularly in South Europe where infrastructure is particularly unreliable.

Except for the potential connections with Spain, Noor will not integrate with the European grid architecture, meaning there will be little or no direct impact on the stability of the European energy system. The possibility of future development would be based on market demand, but it would also jeopardise Noor's focus on local distribution, making it susceptible to accusations of neocolonialism that have plagued DESERTEC.

Within its closed distribution network, however, Noor does promote increased stability through its multiple facilities, reducing the potential impact of sudden disruptions.

ii. Flexibility

Concentrated solar power plants have not proven to be highly flexible in terms of balancing short-term grid imbalances, but do have reserves in the form of thermal storage for short term (minutes-to-hours) added generation. Therefore, the proposed DESERTEC concept does provide some added flexibility, both locally and for Europe as a whole. At the same time, the radial construction of the DESERTEC transmission cables would not substantially improve the system's ability to respond to specific interruptions as convincingly as to general issues.

Noor, on the other hand, will improve system flexibility in Morocco due primarily to the short distance from production to consumer and the integrated mesh distribution construction of the grid. Increasing the source options, as Noor does, is essential to a flexible overall system.

iii. Resilience

The Noor project is considerably less vulnerable to externalities, such as political unrest outside Morocco's borders or meteorological, geological and climate uncertainties, than the existing import-dependent energy system. The financing structure has also contributed to the resilience of the project by creating partnerships with international organisations and creditors that will work to ensure the continued success of Noor.

The DESERTEC concept should improve Europe's energy system resilience due to the diversification of supply. However, it will come as no surprise that resilience is DESERTEC's most glaring deficit.

The vulnerability of transmission lines crossing MENA, then the Mediterranean Sea, and arriving in Europe is clear. Although the concept was developed prior to the Arab Spring, it is hard to imagine the geopolitical assumptions necessary to construct a system dependent on the cooperation of dictatorial regimes such as Muammar Gaddafi's Libya. Political instability is not specific to MENA, but the history of European colonialism in the region and continued military activities there make increasing dependency on the region suspect and unadvisable. Certainly there is the hope that a concept such as DESERTEC could play a role in establishing a productive, interdependent relationship between the EU and MENA, leading to more stability, but if the objective is energy security, DESERTEC is, at best, ahead of its time.

iv. **Adequacy**

Solar power has the significant advantage to be productive during peak load hours around midday. On the other hand, only limited added generation is available in the evening and at night time. Further, CSP relies on direct insolation. An adverse weather situation with dense cloud coverage and low yield, coupled with high demand and possibly a failure of supply in Europe would soon bring the system to its limits.

Noor, as part of an integrated local system, improves adequacy for the Moroccan energy system. Its impact on the European system is negligible. DESERTEC would only further support adequacy of the European system, though questions surrounding the full market integration between MENA and the EU remain. As previously mentioned, MENA is a very diverse area with actors playing varying roles in the global energy market. With geopolitical and security issues prevalent, it is unlikely that full market integration would survive volatile price fluctuations.

Adequacy requires not only certainty of supply but certainty of supply at a fair market price. The full energy market integration required by the DESERTEC concept would increase European dependency on MENA, thereby reducing affordable adequacy and the system's security.

v. **Robustness**

Establishing long-term co-dependencies with unstable governing regimes in the MENA region undermines European values and limits Europe's political independence. Europe reduces its market choice by engaging in any form of DESERTEC-oriented approach.

Growing demand in the MENA region could sooner or later eradicate any gains for Europe as the generating regions will have a natural priority access to their own generated electricity, regardless of conceptual claims, and would curtail European customers. On the other hand, purchasing power may be higher in Europe, thus driving the much-needed electricity away from the generating countries and possibly resulting in political and market instability.

The Noor project addresses both of these issues through its focus on local markets. It is clear that Morocco views Noor as not just a step toward increased energy security, but as an opportunity for social and economic investment, making it a cornerstone of a robust vision.

5.2.4 Assessing the Potential Contribution to Low-Carbon Energy Security

The previous assessment and analysis has shown that DESERTEC or CSP in general has very high technical potential in the MENA region and can contribute to climate change mitigation. However, the negative energy security implications for Europe as well as the potential geopolitical fallout of an enforced energy co-dependency between Europe and the MENA region prohibits anyone from entertaining the idea of actually implementing an EU–MENA energy integration.

CSP will contribute to low-carbon energy security in MENA and thus indirectly also support security and energy security in Europe.

However, the DESERTEC case is a prime example of the need to apply energy responsibility which implies domestic energy sufficiency.

The MENA projects show that localised and regionalised approaches within Europe will offer better contributions to low-carbon energy security than an expansion of the Energy Union beyond the Mediterranean without preceding political processes.

While the Union for the Mediterranean is certainly meant as being the core of such a process, it is by far insufficient in many dimensions. The current geopolitical shape of MENA renders any such approach infeasible in the time horizon of the next 20 years.

As such, Europe needs to achieve low-carbon energy security from within, as Morocco has begun to implement with Noor.

This does not preclude energy cooperation and supporting the development of local renewable energy markets in the MENA countries with technology, financing and expertise. Still, Europeans also need to be aware of the colonial history in the region and resulting resentments against anything bordering on European influence or control.

5.3 THE WINDS OF THE NORTH SEA

Much like conventional energy sources, accessible supplies of RES are generally concentrated in specific geographical areas without regard to territorial or political boundaries, often motivating the construction of regional large-scale projects to maximise the productivity of the source while minimising the costs of resource extraction or collection. Where RES are concerned, however, these areas of dense resource concentration frequently entail extreme meteorological conditions – such as intense sun, and therefore heat, or high winds – which make them either uninhabitable or, at least, physically removed from major centres of consumption, requiring consideration of further issues surrounding storage, transmission and security. Despite these challenges, however, large-scale offshore wind parks have generated increasing interest over the past decade, and several national governments are coordinating efforts to use the winds of the North and Baltic Seas to produce a more secure energy supply for Europe. In this chapter, we discuss the North Seas Countries Offshore Grid Initiative (NSCOGI) and the Kriegers Flak project in the Baltic Sea to assess their energy security implications and whether large-scale offshore wind farms are a viable solution in creating a low-carbon secure energy future for Europe.

Offshore wind energy developments in European waters have been on a decisively progressive trajectory since 2005 (Fig. 5.2), although their overall contribution to composite renewable energy generation in Europe remains relatively small (European Wind Energy Association, 2015). Current and continued activity patterns focus on the northern coastal countries. In just the first

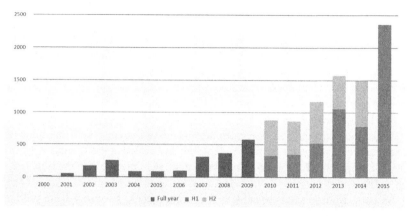

FIGURE 5.2 Annual installed offshore wind capacity in Europe (MW). *EWEA.*

half of 2015, projects encompassing Germany, Denmark, the Netherlands and the United Kingdom accounted for 2340 MW of capacity fully connected to the grid. Of that newly installed capacity, more than 50% are controlled by six developers: RWE (12.3%), E.ON (11.5%), DONG (9.3%), Stadtwerke München (9.1%), Trianel (8.3%) and Energie Baden-Württemberg (EnBW; 5.6%). In the United Kingdom and Germany alone, investments in offshore wind energy totalled more than €7 billion between January and June 2015, and projections through the end of 2016 foresee an additional €10 billion. These numbers can certainly be seen as encouraging, particularly when considered as an initial phase of a broader transition to RES, but the exclusivity of stakeholder participation brings into question the benefits the current course will have on regional and European-wide energy security.

Centralisation and consolidation can be seen as a prudent aspect of the opening stages of the low-carbon energy security transition in Europe; restructuring the entire European grid, for example, is neither financially affordable nor practically tenable. At the same time, however, it continues a cycle of dependency, whether on nations, corporations or processes, fundamentally anathema to a democratised energy supply. Offshore wind projects do benefit from both foundational technology enhancements that are simultaneously progressive and dependable and an integrative infrastructure that reduces, though does not completely allay, implementation and distribution concerns. Though technical specifications and engineering aspects of these projects are essential factors in evaluating their potential contribution to the overall energy supply, the objective of this chapter, more specifically, is to assess the contributions to energy security. To that end, the NSCOGI and Kriegers Flak projects have been selected under the assumption that each provides maximised or best-possible technical potentialities for large-scale offshore wind projects, allowing for the analysis of the projects from a policy perspective. Therefore, the following sections will first review the overall energy policy framework within which these initiatives operate. Second, the constitutive properties of the projects will be discussed, providing an opportunity to evaluate the potential for replicability or scalability. Third, the five-point framework for assessing the implications for energy security will be applied. Finally, the potential contributions of large-scale offshore wind projects to low-carbon energy security will be presented.

A final conclusion will provide opportunities to continue questioning and improving the policy approaches to integrating low-carbon energy sources into a comprehensively secure energy supply.

5.3.1 Energy Policies and Support Mechanisms Around the North and Baltic Seas

The policy umbrella for multinational energy projects within the European Union is set by European energy policies, most notably the Energy Union (European Commission, 2014a). The Energy Union consists of five dimensions:

- First, the Energy Union aims at supply security by diversifying sources of energy and improving the efficiency of the generation and distribution of energy within the EU.
- Second, it promotes a fully integrated internal energy market, enabling energy to flow across the EU without any technical or regulatory barriers, allowing energy providers to compete and provide competitive energy prices.
- Third, the Energy Union fosters energy efficiency in order to reduce pollution and reduce the consumption of domestic and imported energy.
- Fourth, it supports greenhouse gas (GHG) emissions reductions by renewing the EU Emissions Trading System, supporting a global agreement to address climate change through the United Nations Framework Convention on Climate Change and encouraging private investment in low-carbon infrastructure and technologies.
- Fifth, it supports research and innovation in low-carbon technologies by coordinating research and public-private-partnerships.

The priorities set forth by the Energy Union have a distinctly supply-side orientation, with strategies focused heavily on integrating market-based measures and the expansion of infrastructure. Notably, the consumption reduction agenda relies primarily on streamlining efficiencies in generation and distribution. As for energy markets, NordPool, the world's largest multinational exchange trading electrical power, is recognised as a viable vehicle for further integration, though clear deficits in the Baltic States and south-east Europe remain (European Commission, 2014a, p. 4).

Expanding and integrating the disparate infrastructure on a multinational level present considerable financial and policy barriers. The European Energy

Security Strategy (European Commission, 2014b, p. 9) estimates the cost of achieving the required energy transmission infrastructure, in particular cross-border interconnections between Member States, at €200 billion, but such a price tag is seen as twice that which private investors, including the energy sector, would be willing or able to invest. In order to overcome these barriers and motivate the supplementation of this gap in funding, the European Commission identified a set of six electricity infrastructure projects to be implemented with priority, designating the majority of them projects of common interest (PCI) under the Trans-European Networks – Energy (TEN-E) guidelines. In early 2014, the European Council suggested an accelerated implementation of measures aimed at achieving a 10% interconnection target and even suggested increasing the target to 15% by 2030 (European Commission, 2014a, p. 5).

Yet despite general consensus and political determination, financial support for the projects, including the PCIs, has been unable to surpass €5.8 billion, less than 3% of the total identified funding required (European Commission, 2014b, p. 9). There are significant questions as to the ability, or willingness, of Member States and private industry to mobilise the remaining sums necessary for project implementation, particularly in times of national austerity regimes and slow and uncertain economic growth. Moreover, the considerable attention given natural gas interconnectors may prove a further hindrance in terms of spurring investment in infrastructure integration investments. As electricity continues to increase its share of heat and transport energy demand, as well as provide a more viable path toward energy independence, a crucial tenet of energy security, it is arguably the electricity infrastructure that will be key to realising the integration goals of the Energy Union.

Member States' individual governance structures are additional factors in determining or hindering the viability of integrated energy infrastructure initiatives in Europe. As was reported by the director of the European Regional Development Fund (Derdevet, 2015, p. 18), differences in the national energy transmission governance, specifically the mission description for the transmission system operators (TSOs) are identified as obstacles for a better integration and interconnection of the various national energy systems, leading to inertia and suboptimal investments, eventually impacting the energy price and, thereby, the sustainability of a market-based, multinational system.

Two Member States of particular relevance for the NSCOGI and Kriegers Flak project in the Baltic Sea are Germany and Denmark. Both countries have begun pursuing ambitious energy transition agendas with the specified targets shown in Table 5.1 (Federal Ministry for Economic Affairs and Energy, 2015; The Danish Government, 2011).

Compared to Germany's gradual transition, which arguably allows for augmentation and adjustments to processes, the Danish model is focused specifically on end-usage, whether of electricity in 2020 or gross energy in 2050. Whereas Denmark will ostensibly be 100% renewable by 2050, Germany may actually use less total energy (relative to demographic metrics or GDP).

TABLE 5.1 Targets for the Energy Transition in Denmark and Germany

	Germany	Denmark
Share of renewable energy in electricity	50% by 2030	50% by 2020
	65% by 2045	
	80% by 2050	
Share of renewable energy in gross energy consumption	30% by 2030	100% by 2050
	45% by 2040	
	60% by 2050	

The German energy transition, branded the *Energiewende*, is founded on a very specific architecture (German Federal Government, 2010):

1. climate mitigation target: 40% reduction in GHG emissions from 1990 levels by 2020
2. nuclear phase out by 2022
3. competitiveness
4. energy security

The Danish concept, in contrast, focuses almost exclusively on the decarbonisation of the energy system, and considers security of supply and competitiveness as boundary conditions, rather than as stand-alone objectives (The Danish Government, 2011). It becomes apparent that the energy visions of Denmark and Germany differ not only in terms of statistical targets, but also with regard to overall energy policy, including energy security.

Both countries offer various differentiated support mechanisms, primarily in the form of feed-in tariffs and increasingly in the form of market-based support via auctioning. Denmark led this development with opening the tendering for offshore wind energy in 2013 (Danish Energy Agency, 2013). Germany followed with its reform of the renewable energy act in 2015, instituting auctions first for ground-mounted PV, to be expanded into on- and offshore wind (Federal Ministry for Economic Affairs and Energy, 2014; German Federal Government, 2014). Whether such national structures are themselves scalable to a regional or Europe-wide integration architecture remains to be seen.

5.3.2 North Seas Offshore Grid and Kriegers Flak Project Backgrounds

NSCOGI and Kriegers Flak are designated PCIs with priority implementation. While Kriegers Flak is a much more confined project comprising a single wind park and two countries, NSCOGI entails a larger overall footprint.

The idea for a North Seas Offshore Grid materialised in 2010 with the signing of a memorandum of understanding between the 10 neighbouring Member States around the North Seas, their TSOs, the European Network of Transmission System Operators for Electricity (ENTSO-E), regulators represented by the Agency for the Cooperation of Energy Regulators (ACER) and the European Commission (NSCOGI, 2010).

The parts of NSCOGI currently under development are (PLATTS, 2015):

- Norway–United Kingdom interconnection (PCI 1.10), a 1400MW cable to be commissioned in 2020;
- Germany–Norway interconnection between Wilster (DE) and Tonstad (NO) (a.k.a. NORD.LINK project; PCI 1.8), a 1400MW 500kV cable with 520–600km length to be commissioned in 2018;
- Denmark–Netherlands interconnection between Endrup (DK) and Eemshaven (NL) (PCI 1.5), a 700MW 320kV high voltage direct current (HVDC) link of 350km connecting new offshore wind farms to be commissioned in 2019;
- France–Ireland interconnection between La Martyre (FR) and Great Island or Knockraha (IE) (PCI 1.6), a 700MW 320–500kV 600km cable to be commissioned in 2025;
- France–United Kingdom interconnection between Coquelles (FR) and Folkestone (UK) (currently known as the ElecLink project) (PCI 1.7.3), a 1000MW 320kV 51km cable to be commissioned in 2016;
- PCI 1.9 cluster connecting generation from RES in Ireland to the United Kingdom, commissioning estimated around 2020;
- Belgium, two grid-ready offshore hubs connected to the onshore substation Zeebrugge (BE) with anticipatory investments enabling future interconnections with France and/or UK (PCI 1.2) approximately 35km to be commissioned in 2015.

There exist several more proposed projects to further the scope of NSCOGI, though none of these has been finalised, nor is further implementation of additional projects expected before 2017, with commissioning unlikely before 2020 (Fig. 5.3).

In terms of integrating infrastructure and gross energy distribution potential, the current schematic undoubtedly improves interconnectivity, one of the stated goals of the Energy Union. However, the point-to-point, bilateral connections also call into the question the efficiency of the blueprint. It would seem that a more systems-oriented, mesh design would streamline the number of cables and allow for clustering wind farms on the high seas, saving both time and money (Cole et al., 2014). Understandably, a mesh or web approach would require a much higher level of coordination among the participating Member States and the European Commission, specifically ENTSO-E, but it would offer considerable trade-off benefits, including reduced environmental impact and improved systems efficiency of the grid itself.

While a general meshed approach may require too much coordination, simple integration at the local level may offer a viable alternative to both the

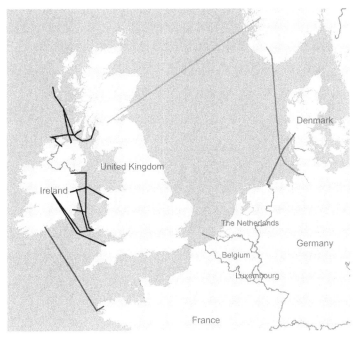

FIGURE 5.3 Northern sea projects of common interest. *Mapbox, OpenStreetMap for the administrative boundaries European Commission for the underlying energy network; Cartography: Ecolgic Instititute.*

point-to-point (radial) and the fully meshed (hub) approach by simply integrating offshore wind energy into the transmission network (NorthSeaGrid, 2015). The NorthSeaGrid project found that key barriers to integration include:

- cross-border integration of renewable energy support schemes: currently Member States restrict their support to domestic renewable generation only, based on the exclusive economic zone (EEZ);
- grid-access responsibility: currently Member States connect to wind farms in their EEZ, even if this is less efficient that a cross-border connection;
- inflexible connection design: connections for wind parks are planned separately from interconnections which can result in stranded assets;
- mismatches in grid connection: if a wind park is planned to be connected to two Member States but one party delays or obstructs the connection, the wind park cannot operate efficiently; and
- differences in priority feed-in and curtailment compensation.

Currently, these regulatory and/or governance barriers are not addressed through the memorandum of understanding or in active policy negotiations among the Member States. Integration, in this sense, is apparently more a matter of increasing supply to or from a certain location rather than a balancing of

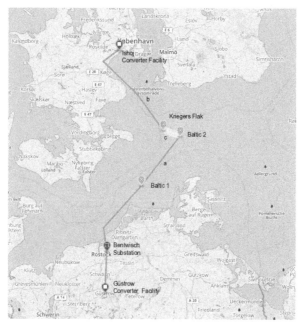

FIGURE 5.4 Kriegers Flak combined grid Danish Energinet.dk connection. *Mapbox, OpenStreetMap for the administrative boundaries Energinet.dk for the underlying energy network; Cartography: Ecologic Institute.*

distribution based on demand. Nevertheless, the number of planned connections in the North Sea will de facto change the European energy landscape, first by enabling increased integration between the Northern energy systems and, second, by opening access for offshore wind energy development along the transmission corridors.

In a sense, NSCOGI is a multiplication, if not a scaled-up replication, of Kriegers Flak in the Baltic Sea. The Kriegers Flak project combines an offshore wind park with interconnected transmission across Denmark and Germany, an example of the radial layout of NSCOGI (Fig. 5.4). By investigating Kriegers Flak, it may be possible to gain insight into the possibilities and limitations of the NSCOGI plan.

Kriegers Flak is a project by the Danish Energinet.dk in collaboration with the German 50 Hz TSO to build the world's first offshore power grid, connecting wind turbines in the Baltic Sea with the two EEZs of Denmark and Germany (Denmark.dk, n.d.). The project will not only enable renewable power generation with a planned capacity of 600 MW, but also improve the interconnection between Germany and Denmark. The two TSOs collectively manage the grid connection, while the Danish Energy Agency is tendering the wind concession in the Danish EEZ.

The Kriegers Flak interconnection is known as the Kriegers Flak Combined Grid Solution PCI 4.1 and runs between Ishőj/Bjæverskov (DK) and Bentwisch/Güstrow

(DE) via offshore wind parks Kriegers Flak (DK) and Baltic 2 (DE) (PLATTS, 2015). The Danish Energy Agency published the tender conditions in November 2015 and expects a final decision on the wind farm tender by the end of 2016, followed by completion of the wind park by 2022 (Danish Energy Agency, 2015b).

Seven potential investors have been prequalified as tenderers for the process, four of which are wholly or substantially controlled by companies previously mentioned in this chapter for their influence in already existing offshore wind projects and two of which were not established prior to the announcement (Danish Energy Agency, 2015a):

1. Kriegers Flak ApS (owned by EnBW AG);
2. Wpd HOFOR Stadtwerke München, Kriegers Flak SPV;
3. European Energy Offshore Consortium (owned by European Energy A/S and Boralex Europe SA);
4. A not-yet-established company by Vattenfall Vindkraft A/S;
5. Kriegers Flak Offshore Wind I/S (a not yet established company by Statoil ASA and E.ON Wind Denmark AB);
6. ScottishPower Renewable Energy Limited; and
7. DONG Energy Wind Power A/S.

Since September 2015, the German Baltic 2 Park, currently fully owned by the German utility EnBW, operates adjacent to the Danish wind park. Though considerably smaller with 228 MW, it provides a connection to the German mainland grid (EnBW, 2015). The commissioning date for the 600 MW 320 kV 270 km long grid interconnection is expected for 2018 (PLATTS, 2015).

Both NSCOGI and Kriegers Flak projects assume directly connecting mainland Europe to hydropower pump-storage in Scandinavia. Norway and Sweden were determined to have a combined hydrostorage capacity of 35 GW in 2010 (Ess et al., 2012). Pump-storage, however, is only a small fraction, less than 1.5 GW, of total hydrostorage capacity, of which more than 90% is in Norway. Norway is considering adding significant additional pump-storage capacity, up to 20 GW (Solvang et al., 2014). The majority of the added capacity could derive from existing hydroplants as much as 18 GW from 12 plants, thus reducing the environmental impact and costs. Transmission connection to Western Europe would be considerably less expensive in Norway compared to Sweden where new pump-storage would be planned for the remote North. However, even in the more accessible Norwegian scenario, the full 20 GW would only become available in 2030, assuming immediate construction of 1 GW added pump-storage capacity per year through 2020 and even higher rates thereafter. Total interconnection between Norway and Member States is expected to more than double from the current 4.8–11.5 GW in 2020, assuming key PCI steps are implemented.

Considering the enormous costs and logistical challenges of this large-scale peripheral project, there remains significant uncertainty regarding the time schedule and the actual added pump-storage capacity by the time the new PCI interconnectors become available, leading to further doubts for potential investors and the timeframe for economic benefits for consumers. Still, pump-storage is the only available economically feasible storage technology today. While costs are a key driver in any infrastructure development, even more crucial is the unmatched flexibility pump-storage hydro offers (Eurelectric, 2011): with around 0.1 h start-up time, they are online faster than any other balancing power which range between 1.5 h for combined-cycle gas power plants to 40 h for nuclear plants, and offering nominal load-in or decreases of up to 40% per minute, compared to 2–4% for other plants.

5.3.3 Energy Security Implications

Due to the simple fact that NSCOGI and Kriegers Flak are still in development, assessing the energy security implications is an exercise in forecasting and application of any such assessment to the broader category of large-scale offshore wind energy projects is incomplete. Of course, this is in and of itself an issue regarding the effectiveness of these projects to provide increased energy security: planning and implementation proceed over an exaggerated and unreliable timeline, making their benefits unrealisable within current realities. Nonetheless, there is ample information to evaluate the constitutive policies that have led to the creation of these projects. Bearing this in mind and applying the five-point framework previously outlined, it is possible to explore the potential contribution of both NSCOGI and its smaller forerunner Kriegers Flak in the Baltic Sea to European low-carbon energy security.

i. **Stability**

Theoretically, diversification is the most direct way to improve stability of the energy grid and support energy security. Because NSCOGI and Kriegers Flak work within a larger energy paradigm, there is reason to believe that the addition of offshore generated wind energy would improve the stability of the electrical grids at local, national and regional levels. At the same time, it would be prudent to raise concerns as to the reliability, a subcategory of stability, of below-sea interconnections, due to the simple fact that they remain untested. The long distances between connection points raise caution regarding the balancing potential for short-term imbalances.

Under this criterion, Kriegers Flak would provide more stability to the broader grid system. In the point-to-point context, NSCOGI would also provide greater stability. However, in the aggregate, the project cannot be seen to essentially improve stability throughout its footprint. For example, a disruption in Germany would not be ameliorated by the France–Ireland interconnector. Although the potential for stability is high, the project's design is primarily supply oriented, whereas stability disruptions are transmission or usage oriented.

ii. **Flexibility**

The proposed interconnections certainly do improve the interconnection capacity and thus enable power balancing across borders, but it would also provide access to the pump storage balancing capacity in the Scandinavian countries, especially Norway. Given the extremely fast response-times of pump-storage, grid flexibility would be significantly improved throughout the interconnected electricity grids. In addition, interconnected offshore wind parks, such as at Kriegers Flak, will further improve the system flexibility, provided they can curtail generation.

A major challenge in the North Seas Offshore Grid and the Kriegers Flak case is the closeness of consumption patterns and, thus, limited difference in peak and off-peak consumption. Moreover, weather patterns are not sufficiently different to ensure alternating wind potential: when high winds blow in the Channel, they will likely also translate into high winds in the rest of the North Sea, resulting in high wind output in the entire NSCOGI area. Some demand balancing may result from the 1 h time zone difference between Ireland and the United Kingdom and the remaining eight Member States.

Therefore, offshore wind energy does provide the potential for significant increased flexibility of the broader grid. However, the geographic concentration of the generation area reduces the likelihood that this would be of great benefit to the larger question of energy security.

iii. Resilience

NSCOGI and Kriegers Flak will both significantly improve grid resilience, due primarily to interconnectors. In case of disruption of one or few interconnectors, neighbouring interconnectors can take up a part of the transmission load, provided sufficient reserves exist.

Moreover, connecting Kriegers Flak to both Denmark and Germany allows for added resilience compared to a mono-connection. NSCOGI's design, in this sense, also benefits resilience of the grid architecture. However, capacity levels as proposed appear to be inadequate to provide consistent assurances to longer-term usage requirements. It is difficult to see how the current iterations of interconnectors can provide resilience on a longer time horizon.

Still, compared to the current models, system resilience will be greatly improved.

iv. Adequacy

Adequacy takes into consideration the question of the entire energy system to meet demand at all times under all anticipated conditions. Again, as with stability, diversification is essential to this criterion, which offshore wind energy projects would appear to support. However, in contrast to the stability assessment, it is not the source diversity that is paramount, but transmission diversity, when evaluating adequacy. In that sense, NSCOGI and Kriegers Flak do not improve system adequacy.

There are three dominant reasons for this assessment. First, the radial design of the interconnectors limits the ability of the markets to adapt to demand variations. Second, the dependency on hydropump storage in Norway and Sweden expand, rather than reduce, potential vulnerabilities to transmission. And third, the homogeneity of the markets served by the projects undermines viable opportunities for balancing and rerouting.

Although Kriegers Flak and NSCOGI are sheltered from many of the market disruptors the current energy paradigm faces, they do not reduce the potential for new disruptors, thereby failing to improve the adequacy of the overall energy system.

v. **Robustness**

As stipulated in the European Commission's 2015 summer package (European Commission, 2015), choice for consumers will increase in the future Energy Union. NSCOGI and Kriegers Flak, just as all other PCIs, play a role in establishing this level playing field and empowering consumers via new market design. However, the specific contribution of the planned interconnected offshore wind parks will be small compared to other actions such as improved consumption and cost information, eased switching of supplier, dynamic tariffs, collective and community schemes and smart homes (European Commission, 2015).

Robustness requires a fully realised integration of the market. Physical interconnections are a necessary condition for enabling the full market integration towards the European Internal Electricity Market, enabling cross-border trading and pricing across the entire European Union (European Commission, 2014a). While day-ahead price-coupling already includes 17 Member States in south-western and north-western Europe, intraday trading is not yet coupled (ENTSOE, 2014). The added interconnections promise to become the back bone for a fully integrated European energy-only market as aspired to by the Energy Union.

Yet as has been discussed in this chapter, the current support mechanisms for spurring investment and private sector participation in these large-scale projects have not resulted in a consumer-oriented level playing field. Very few actors are participating, and this will lead to a cartel-like system with control of the energy supply in very few hands. In order to create a robust energy paradigm, incentives must lead to a democratisation of the energy sector, which the current projects do not.

5.3.4 Assessing the Potential Contribution to Low-Carbon Energy Security

NSCOGI and Kriegers Flak are well-intentioned efforts to reduce the carbon intensity of the energy supply in Europe. To that end, it can be said that they are positive steps in the transition to a low-carbon energy future. Unfortunately, these projects do not take advantage of the opportunity to shift the energy system paradigm, to rethink the manner and method of collecting and distributing energy in Europe, a rethinking necessary to provide low-carbon energy security.

Much like conventional energy sourcing, these large-scale offshore grid initiatives are predicated on building centralised extraction (or collection) facilities, then distributing energy via a radial connector, like a pipeline, to the consumer. Infrastructure is nearly obsolete before it is commissioned. Technologies advance before the systems are financially self-sustaining. And market expansion places the consumer further and further away from the source, creating a wider discrepancy between market price and production cost. All of this adds up to a reiteration of the current energy paradigm, replacing carbon-intense energy with low-carbon energy.

Such a transition should ostensibly be a step toward a more secure energy system, if only due to a reduced dependence on extra-European imports. However, this paradigm has led Europe to its current status, and maintaining it would not improve energy security. Europe's energy insecurity is fundamentally due to market distortions through inefficient distribution, cartel-like supply chains and inaccurate pricing mechanisms. Lasting low-carbon energy security requires that these issues be addressed. NSCOGI and Kriegers Flak, however, only reify them.

5.4 LESSONS FOR A LOW-CARBON AND SECURE ENERGY TRANSITION

Europe's existing energy security paradigm is founded on the same principles that have coupled energy consumption with economic growth, expanded the map in terms of system interdependencies and relied on existing transmission regimes and impaired the implementation of comprehensive changes, including a total transition to RES. The mega-projects discussed in this chapter were devised and, where possible, implemented under this paradigm that continues to govern policy decisions today. The successes and shortcomings of these initiatives provide us with opportunities to not only assess the technical and logistical potentialities of large-scale offshore wind and solar projects; we can also learn how to rethink the existing energy security paradigm, reshaping equilibria between capacity and efficiency or interdependency and vulnerability in developing low-carbon energy security for Europe.

The mega-projects in MENA offer contrasting approaches to constructing an interdependent energy system. Whereas DESERTEC proposed a continuation of the existing paradigm by designing a system founded on physical dependencies, particularly with transmission lines across the Mediterranean Sea, Morocco's Noor project builds a network of support through financial and technological exchange, enabling further advancements, promoting economic partnerships and stabilising political relationships. By reducing the vulnerabilities inherent to physical dependencies, Noor has provided insight into a new way of increasing security through interdependency.

While traditional energy policies focused predominantly on expanding capacity, including also transmission capacity, novel thinking links generation, transmission and demand via a systems approach, as can be seen to some degree at Kriegers Flak. While arguably omitting any direct integration of demand management to complete the systems approach, Kriegers Flak is the very first integrated offshore wind farm with integration binational interconnection, thus allowing excess electricity to flow either to Denmark or Germany and improving the efficiency, not solely the generating capacity. NSCOGI, on the other hand, has failed to focus on efficient transmission due to its radial design and overcomplication of the interconnection scheme.

Reshaping equilibria for low-carbon energy security with the help of large-scale renewable energy projects leads to much deeper questions of the prevailing energy policy paradigms. As long as interdependency continues to be inhibited by system vulnerabilities and efficiency remains secondary to generated capacity, low-carbon energy security in Europe will be a pipe dream.

REFERENCES

African Bank for Development, 2013. Ouarzazate I Concentrated Solar Power (CSP) Project Fact Sheet. Retrieved from: http://www.afdb.org/fileadmin/uploads/afdb/Documents/Generic-Documents/Ouarzazate%20I%20-%20Fact%20Sheet.pdf.

African Bank for Development, 2014. Summary of Environmental and Social Impact Assessment.

Cole, S., Martinot, P., Rapoport, S., Papaefthymiou, G., Gori, V., 2014. Study of the Benefits of a Meshed Offshore Grid in Northern Seas Region. European Commission.

Crivello, S., Karatayev, M., Lombardi, P., Cotella, G., Santangelo, M. (Eds.), 2013. Deliverable 1.3 Report on Main Trends in European Energy Policies. Retrieved from: http://www.milesecure2050.eu/documents/public-deliverables/en/deliverable-1-3-report-on-main-trends-in-european-energy-policies.

Danish Energy Agency, 2013. New Offshore Wind Tenders in Denmark.

Danish Energy Agency, 28 October 2015a. Great Competition for the Tender of the Offshore Wind Farm Kriegers Flak. Danish Energy Agency. Retrieved from: http://www.ens.dk/en/info/news-danish-energy-agency/great-competition-tender-offshore-wind-farm-kriegers-flak.

Danish Energy Agency, 20 November 2015b. English Translation of Tender Conditions for Kriegers Flak Offshore Wind Farm. Danish Energy Agency. Retrieved from: http://www.ens.dk/sites/ens.dk/files/byggeri/20151009_udbudsvilkar_uk_published_20_11_15.pdf.

Denmark.dk (n.d.). Kriegers Flak – The World's First Offshore Electricity "Supergrid". Retrieved from: http://denmark.dk/en/green-living/sustainable-projects/kriegers-flak-the-worlds-first-offshore-electricity-supergrid/.

Derdevet, M., 2015. Energy, a Networked Europe: Twelve Proposals for a Common Energy Infrastructure Policy. La Documentation Francaise.

DESERTEC Foundation (n.d.). DESERTEC Milestones. Retrieved from: http://www.desertec.org/global-mission/milestones/.

DESERTEC Foundation, 2010. Red Paper – An Overview of the DESERTEC Concept. Retrieved from: http://www.desertec.org/fileadmin/downloads/desertec-foundation_redpaper_3rd-edition_english.pdf.

El Nokrashy, H., 2005. Renewable Energy Partnership Europe – Middle East – North Africa. p. 5 Retrieved from: http://www.menarec.org/resources/RenewableEnergyPartnershipEUMENA_MENAREC2.pdf.

EnBW, 2015. Windkraft in Neuer Dimension. EnBW. Retrieved from: https://www.enbw.com/media/konzern/docs/energieerzeugung/flyer-baltic-2.pdf.

ENTSOE, 2014. South-Western and North-Western Europe Market Coupling Project Go-Live. ENTSOE. Retrieved from: https://www.entsoe.eu/news-events/announcements/announcements-archive/Pages/News/SWE-NWE-Market-Coupling-Go-Live.aspx.

Ess, F., Haefke, L., Hobohm, J., Peter, F., Wünsch, M., 2012. The Significance of International Hydropower Storage for the Energy Transition. Prognos AG. Retrieved from: http://www.prognos.com/uploads/tx_atwpubdb/121023_Prognos_Study_International_storage_EN.pdf.

Euractiv, 9 August 2013. Desertec Abandons Sahara Solar Power Export Dream. Retrieved from: http://www.euractiv.com/energy/desertec-abandons-sahara-solar-p-news-528151.

Eurelectric, 2011. Hydro in Europe: Powering Renewables. Retrieved from: http://www.eurelectric.org/media/26440/hydro_report_final_110926_01-2011-160-0005-01-e.pdf.

European Commission, 2014a. Energy Union Communication SWD (2014) Final. European Commission. Retrieved from: http://ec.europa.eu/energy/sites/ener/files/documents/2014_iem_communication_annex5.pdf.

European Commission, 2014b. European Energy Security Strategy COM (2014) 330 Final. European Commission. Retrieved from: http://eur-lex.europa.eu/legal-content/EN/TXT/PDF/?uri=CELEX:52014DC0330&from=EN.

European Commission, 2015. Delivering a New Deal for Energy Consumers, COM(2015) 339 Final. European Commission. Retrieved from: https://ec.europa.eu/energy/sites/ener/files/documents/1_EN_ACT_part1_v8.pdf.

European Wind Energy Association, 2015. The European Offshore Wind Industry – Key Trends and Statistics 1st Half 2015. Retrieved from: http://www.ewea.org/fileadmin/files/library/publications/statistics/EWEA-European-Offshore-Statistics-H1-2015.pdf.

Federal Ministry for Economic Affairs and Energy, 2014. Photovoltaics, Wind Power and Biomass: The Reforms at a Glance. Retrieved from: http://www.bmwi.de/English/Redaktion/Pdf/eeg-faktenblatt-neuerungen-auf-einen-blick,property=pdf,bereich=bmwi2012,sprache=en,rwb=true.pdf.

Federal Ministry for Economic Affairs and Energy, 2015. Making a Success of the Energy Transition. p. 32. Retrieved from: http://www.bmwi.de/English/Redaktion/Pdf/making-a-success-of-the-energy-transition,property=pdf,bereich=bmwi2012,sprache=en,rwb=true.pdf.

French Government, 13 July 2008. Joint Declaration of the Paris Summit for the Mediterranean. Retrieved from: http://ufmsecretariat.org/wp-content/uploads/2015/10/ufm_paris_declaration1.pdf.

Friedman, L., 20 June 2011. Can North Africa Light Up Europe with Concentrated Solar Power? New York Times. Retrieved from: http://www.nytimes.com/cwire/2011/06/20/20climatewire-can-north-africa-light-up-europe-with-concen-79708.html?pagewanted=all.

German Federal Government, (First), 2010. Energiekonzept für eine umweltschonende, zuverlässige und bezahlbare Energieversorgung. Retrieved from: http://www.bundesregierung.de/ContentArchiv/DE/Archiv17/_Anlagen/2012/02/energiekonzept-final.pdf;jsessionid=915CE03A123028E6A697E477F7E54F58.s2t1?__blob=publicationFile&v=5.

German Federal Government, 2014. Act on the Development of Renewable Energy Sources. Presse- und Informationsamt der Bundesregierung. Retrieved from: http://www.bmwi.de/English/Redaktion/Pdf/renewable-energy-sources-act-eeg-2014,property=pdf,bereich=bmwi2012,sprache=en,rwb=true.pdf.

Hamouchene, H., March 2015. Desertec: The Renewable Energy Grab? New Internationalist. Retrieved from: http://newint.org/features/2015/03/01/desertec-long/.

Hickman, L., 2 November 2011a. Morocco to Host First Solar Farm in €400bn Renewables Network. The Guardian. Retrieved from: http://www.theguardian.com/environment/2011/nov/02/morocco-solar-farm-renewables.

Hickman, L., 11 December 2011b. Could the Desert Sun Power the World? The Guardian. Retrieved from: http://www.theguardian.com/environment/2011/dec/11/sahara-solar-panels-green-electricity.

Hinnebusch, R., 2012. Europe and the Middle East: from imperialism to liberal peace? Review of European Studies 4 (3), 18–31.

IRENA, 2013. MENA Renewables Status Report. Retrieved from: http://www.ren21.net/REN-21Activities/RegionalStatusReports.aspx.

Kabariti, M., Möller, U., Knies, G., 2003. Trans-Mediterranean Renewable Energy Cooperation "TREC" for development, climate stabilisation and good neighbourhood. In: Presented at the Arab Thought Forum and Club of Rome, Amman Retrieved from: www.desertec.org/downloads/ammanpaper_14102003.pd.

Kwasnieswski, N., 3 April 2013. Desertec on the Ropes: Competitors and Opponents Threaten Energy Plan. Der Spiegel. Retrieved from: http://www.spiegel.de/international/europe/competitors-and-local-opposition-threatens-desertec-solar-plan-a-892332.html.

MED-EMIP, 2010. Overview of the Power Systems of the Mediterranean Basin. Retrieved from: http://www.medemip.eu/Calc/FM/MED-EMIP/MEDRING_Study_Update/Final_Draft/MEDRING-Update-Volume-I-FINAL_Draft_April2010.pdf.

Mobarek, S., 2015. Morocco – MA- Noor Ouarzazate Concentrated Solar Power Project: P131256-Implementation Status Results Report: Sequence 02. Retrieved from: http://www-wds.worldbank.org/external/default/WDSContentServer/WDSP/MNA/2015/11/16/090224b0831beb01/1_0/Rendered/PDF/Morocco000MA000Report000Sequence002.pdf.

NorthSeaGrid, 2015. NorthSeaGrid Integrated Offshore Grid Solutions in the North Sea. Retrieved from: http://www.northseagrid.info/sites/default/files/NorthSeaGrid_SynthesisOfFindings.pdf.

NSCOGI, 2010. The North Seas Countries' Offshore Grid Initiative Memorandum of Understanding. Retrieved from: http://www.benelux.int/download_file/view/3308/3108/.

PLATTS, 2015. Projects of Common Interest. European Commission. Retrieved from: http://ec.europa.eu/energy/infrastructure/transparency_platform/map-viewer/.

Reuters, 19 January 2015. Saudi Arabia's Nuclear, Renewable Energy Plans Pushed Back. Reuters. Retrieved from: http://www.reuters.com/article/2015/01/19/saudi-nuclear-energy-idUSL-6N0UY2LS20150119.

Solvang, E., Charmasson, J., Suaterleute, J., Harby, A., Killingtveit, A., Andersen, O., Aas, O., 2014. Norwegian Hydropower for Large-Scale Electricity Balancing Needs (SINTEF). Retrieved from: http://www.cedren.no/LinkClick.aspx?fileticket=4404a_u9jbc%3d&tabid=4552&portalid=48&mid=5869.

Terrapon-Pfaff, J., Borbonus, S., Viebahn, P., Fink, T., Brand, B., Schinke, B., 2015. Social CSP – Energy and Development: Exploring the Local Livelihood Dimension of the Nooro I CSP Project in Southern Morocco. Retrieved from: https://germanwatch.org/en/download/11797.pdf.

The Danish Government, 2011. Our Future Energy. Retrieved from: http://www.ens.dk/sites/ens.dk/files/policy/danish-climate-energy-policy/our_future_energy.pdf.

Zickfeld, F., Wieland, A., Pudlik, M., Ragwitz, M., Sensfuß, F., 2012. 2050 Desert Power – Perspectives on a Sustainable Power System for EUMENA. Dii and the Fraunhofer Institute for Solar Energy Systems. Retrieved from: http://new.desertenergy.org/wp-content/uploads/2015/09/dp2050_study_web.pdf.

Chapter 6

European Distributed Renewable Energy Case Studies

G. Quinti[1], G. Caiati[1], M. Gruenig[2,3], B. O'Donnell[3], O. Amerighi[4], B. Baldissara[4], B. Felici[4]

[1]Laboratorio di Scienze della Cittadinanza (LSC), Rome, Italy; [2]Ecologic Institute, Berlin, Germany; [3]Ecologic Institute US, Washington, DC, United States; [4]Agenzia nazionale per le nuove tecnologie, l'energia e lo sviluppo economico sostenibile (ENEA), Rome, Italy

6.1 INTRODUCTION

This chapter deals with the importance of renewable energy in Europe. If completing the Energy Union remains one of the top priorities of the European Commission (but not at all a reached goal) and a critical component in Europe's transition towards the decarbonised energy system of the future, and if the European Commission is far from having the authority to set the national energy policies for each of its Member States, then progress in the use of renewable energy can be achieved thanks to the European policies on energy issues.

This chapter will try to illustrate the progress made following three paths:

1. We will investigate on the 'weight' and the 'characteristics' of renewable energy in the frame of the so-called anticipatory experiences (AEs) of energy transition, ie, some case studies at the local level (in Europe) where the transition towards a low-carbon society (or, at least, some of its features) was already implemented (Section 6.2).

2. We will illustrate some national experience with renewable energy, taking into account two of the main important European countries: Germany (the leading European country at the demographic, economic and political level) and Italy (the most important South European country); both countries, albeit with some different choices, ascribe to the renewable energy a considerable importance (Section 6.3 is devoted to Germany and Section 6.4 is devoted to Italy);

3. In a transversal way, we'll address the lessons learned from a decade's worth of European Union (EU) energy transition efforts, highlighting how

climate change mitigation strategies, energy costs and market-oriented policies will create a new set of implications for EU energy security (Section 6.5).

This chapter (as well as the whole book) has been written by many authors (all belonging to the partners' staff of the MILESECURE-2050 project).

6.2 ANTICIPATORY EXPERIENCES: REFLECTING HOW WE ENVISAGE THE LONG-TERM WAYS RENEWABLE ENERGY WILL COMPETE WITH FOSSIL FUELS IN A LIBERALISED MARKET ENVIRONMENT

6.2.1 Anticipatory Experiences in the Energy Transition: What Is It About?

Among the activities pursued through the research project that forms the basis of this book, a pivotal role is played by the 'analysis of concrete anticipatory experiences (AE) on energy transition at the local level'. In more detail, this exercise consisted of the thorough assessment of a large number of concrete local experiences disseminated throughout Europe, where the transition towards a low-carbon society (or, at least, some of its features) was already implemented punctually. The chosen case studies present specific characteristics when compared to other sustainable energy initiatives (namely, operationality, social impact, transparency and systematicity), so as to identify societal conditions that facilitate (or, conversely, hinder) a transition towards a postcarbon (or low-carbon) society.

Research, therefore, focused on a group of initiatives undertaken in Europe to introduce and spread eco-sustainable forms of energy in certain geographical areas. These experiences have been selected because they represent an 'anticipation' of energy transition and, in this sense, are to be considered as existing 'bits' or 'pieces' of a future low-carbon society.

The AEs taken into account in this study have the following four fundamental characteristics, on the basis of which they were selected:

- **Operationality**. The AEs have an operational character, since they involve not only the activation of new technologies (completely new or, more often, new in the local context) but also new forms of organisation for the management of new technologies. Experiences that do not include this element and which are limited, for example, to aspects like training, awareness raising or information, are not anticipatory.
- **Social impact**. Compared to other initiatives, AEs have a significant social impact (producing, for example, changes in common behaviours, the introduction of new rules or regulations, changes in the way people or organisations use resources, changes in the local production system, changes in living and working environments, etc.).

- **Transparency**. AEs are also distinguished by their transparency, having a communicative propensity, providing ready access to information on the results achieved and entering into relationships of interaction and exchange with other initiatives.
- **Systematicity**. AEs have a strong tendency – although not always fully expressed – to assume a systemic character, ie, they operate simultaneously on a plurality of factors involved in energy transition, be they of a technological, economic, social or cultural nature. On the other hand, initiatives that focus on only one technological or social aspect (eg, a single segment of a technology or one type of social behaviour) cannot be considered to be anticipatory.

6.2.2 Anticipatory Experiences in the Energy Transition in Europe

Following these four conditions in the MILESECURE-2050 project mentioned in Section 6.2.1, 90 AE have been identified (from 1500 potential candidates mentioned in various European databases, such as Concerto, Civitas, Intelligent Energy, Global Ecovillage Network Europe, etc. and collected through a call for experiences) in 19 European countries and specifically in the six European countries most populated (France, the United Kingdom, Germany, Italy, Spain and Poland) and in another 13 European countries[1] (Austria, Belgium, Croatia, Denmark, Finland, Hungary, Ireland, Netherlands, Norway, Portugal, Slovenia, Sweden and Switzerland), referring to all European regions (Northern Europe, Central Europe, Eastern Europe and the Mediterranean; Fig. 6.1).

AE are distributed both in the urban and rural contexts, from big cities such as Amsterdam to little islands such as Samsoe (Denmark). Most experiences have been in operation for quite a while (before 1992) and only one-third of those identified were launched after 2005. The completed list of all the AEs with their main features is reported in Table 6.1.

6.2.3 Renewable Energy in Anticipatory Experiences

Several types of activities related to energy transition can be found in the 90 AE identified (sustainable housing, ie, eco-housing and energy savings in buildings; sustainable mobility and transport; sustainable services and industry, ie, actions to reduce greenhouse gas (GHG) emissions in industrial settlements and in services; and renewable energy production, ie, actions to promote solar energy, wind energy and biofuels).

In fact, only 10 from the 90 AE are focused only on renewable energy production.

However, 27 AE more are characterised by a transversal approach, taking into account many types of activities. Renewable energy production/promotion

---

1. By 'European countries' we mean EU member states (EU-28) plus the four EFTA countries (Iceland, Liechtenstein, Norway and Switzerland).

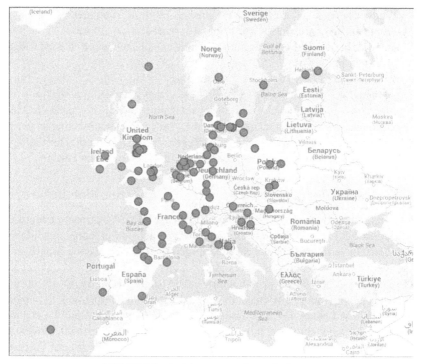

FIGURE 6.1 Anticipatory experiences in Europe.

is taken into account in 20 of them. Therefore, as a whole, we can consider that the issue of renewable energies is in the core of 30 (20 + 10) AE.

More specifically:

- nineteen in promoting solar energy (eg, Barcelona, Spain; Saerbeck, Germany; Daszyna, Poland; Unst Shetland, United Kingdom; Los Molinos del Rio Agua, Spain)
- ten in promoting wind energy (eg, Samsoe, Denmark; Hadyard Ill, United Kingdom; Skive, Denmark)
- nine in promoting bioenergy (eg, Jühnde, Germany; Peccioli, Italy; Kotka, Finland)

This figure is not negligible. However, it represents only less than one-third of the AE identified (ie, renewable energy production is not considered in 60 of the 90 AEs).

6.2.3.1 Renewable Energy in AEs: Some Examples

How is renewable energy taken into account in AE? Some brief case descriptions are reported below. These cases have been chosen among the

TABLE 6.1 List of Anticipatory Experiences

#	Name	City	Country	Sector
1	Superblocks	Vitoria–Gasteiz	ES	Sustainable mobility and transport
2	Lambeth ECOSTILER Community	Borough of Lambeth, London	UK	Sustainable housing
3	Re:FIT London	London	UK	Sustainable services and industry
4	Hadyard Hill Community Energy Project	Hadyard Hill	UK	**Renewable energy production**
5	One Planet Sutton	London Borough of Sutton	UK	Sustainable housing
6	One Brighton	Brighton	UK	Sustainable housing
7	Civitas Bristol	Bristol	UK	Sustainable mobility and transport
8	Civitas Preston	Preston	UK	Sustainable mobility and transport
9	Civitas Ljubljana	Ljubljana	SI	Sustainable mobility and transport
10	Ekostaden Augustenborg	Malmo	SE	Sustainable housing
11	Fossil fuel free – Växjö	Växjö	SE	Transversal sustainable approach
12	Civitas Funchal	Funchal	PT	Sustainable mobility and transport
13	Delft Sesac	Delft	NL	Sustainable housing
14	Noorderplassen-West	Almere	NL	Trasversal sustainable approach
15	Civitas Kracow	Kracow	PL	Sustainable mobility and transport
16	Civitas Szcsecinek	Szcsecinek	PL	Sustainable mobility and transport
17	Zuidbroek Energy Neutral	Apeldoorn	NL	Sustainable housing
18	Leidsche Rijn	Utrecht	NL	Sustainable housing

Continued

TABLE 6.1 List of Anticipatory Experiences—cont'd

#	Name	City	Country	Sector
19	Civitas Stockholm	Stockholm	SE	Sustainable mobility and transport
20	Quartiere Cristo	Alessandria	IT	Sustainable housing
21	Cloughjordan Ecovillage	Cloughjordan	IE	Sustainable housing
22	Civitas Rotterdam	Rotterdam	NL	Sustainable mobility and transport
23	Civitas Zagreb	Zagreb	HR	Sustainable mobility and transport
24	Óbuda Faluház	Budapest	HU	Sustainable housing
25	Zac Pajol	Paris	FR	Trasversal sustainable approach
26	Ile de Nantes District	Nantes	FR	Trasversal sustainable approach
27	Grand Lyon's	Lyon	FR	Trasversal sustainable approach
28	Zac de Bonne	Grenoble	FR	Sustainable housing
29	Waste to Energy in Kotka	Kotka	FI	**Renewable energy production**
30	Lourdes Renove	Tudela	ES	Sustainable housing
31	Antondegi	San Sebastian	ES	Sustainable housing
32	ProjectZero	Sonderborg	DK	Trasversal sustainable approach
33	Civitas Toulouse	Toulouse	FR	Sustainable mobility and transport
34	Solar Ordinance Barcelona	Barcelona	ES	**Renewable energy production**
35	Vauban	Freiburg	DE	Trasversal sustainable approach
36	Samsoe	Samsoe	DK	Trasversal sustainable approach
37	BedZED	London	UK	Trasversal sustainable approach
38	Western Harbour	Malmo	SE	Trasversal sustainable approach

TABLE 6.1 List of Anticipatory Experiences—cont'd

#	Name	City	Country	Sector
39	Eko Vikki	Helsinki	FI	Trasversal sustainable approach
40	Hammarby Sjostad	Stockholm	SE	Trasversal sustainable approach
41	Kronsberg	Hannover	DE	Trasversal sustainable approach
42	London Borough of Lewisham	London	UK	Sustainable housing
43	Sistema Peccioli	Peccioli	IT	**Renewable energy production**
44	Parco Eolico a Varese Ligure	Varese Ligure, Spezia	IT	**Renewable energy production**
45	Civitas Ghent	Ghent	BE	Sustainable mobility and transport
46	Civitas Graz	Graz	AT	Sustainable mobility and transport
47	Scharnhauser Park	Ostfildern	DE	Sustainable housing
48	Emscher Park	Emscher region	DE	Sustainable housing
49	Kirklees Warm Zones	Kirklees	UK	Sustainable housing
50	Eva Lanxmere	Culemborg	NL	Trasversal sustainable approach
51	Bioenergiepark Saerbeck	Saerbeck	DE	**Renewable energy production**
52	Bahnstadt–Heidelberg	Heidelberg	DE	Trasversal sustainable approach
53	EcoMobility in Bremen	Bremen	DE	Sustainable mobility and transport
54	Sustainable Blacon	Blacon	UK	Sustainable housing
55	Ashton Hayes Going Carbon Neutral	Ashton Hayes	UK	Trasversal sustainable approach
56	Baywind Energy Co-operative	Cumbria	UK	**Renewable energy production**

Continued

TABLE 6.1 List of Anticipatory Experiences—cont'd

#	Name	City	Country	Sector
57	Irwell Valley Sustainable Communities Programme	East Salford, Manchester	UK	Sustainable housing
58	Findhorn Ecovillage	Moray	UK	Trasversal sustainable approach
59	Bielsko–Biala	Bielsko–Biala	PL	Trasversal sustainable approach
60	Daszyna	Daszyna	PL	**Renewable energy production**
61	Warsaw Mobility pLAN	Warsaw	PL	Sustainable mobility and transport
62	Kristianstad Fossil Fuel Free Municipality	Kristianstad	DK	Trasversal sustainable approach
63	Kalundborg Eco-Industrial Park	Kalundborg	DK	Sustainable services and industry
64	Copenhagen City of Cyclist	Copenhagen	DK	Sustainable mobility and transport
65	Lehen Sustainable District	Salzburg	AT	Sustainable housing
66	Cernier Solution Project	Cernier	CH	Sustainable housing
67	Frankfurt Passive House Capital	Frankfurt	DE	Sustainable housing
68	Brussels Sustainable Neighbourhoods	Brussels	BE	Trasversal sustainable approach
69	PURE project	Unst, Shetland	UK	Sustainable services and industry
70	Bornholm Bright Green Island	Bornholm	DK	Trasversal sustainable approach
71	Skive Energy Town	Skive	DK	Trasversal sustainable approach
72	Hamburg Climate Policy	Hamburg	DE	Sustainable services and industry
73	Oslo Toll Ring	Oslo	Norway	Sustainable mobility and transport

TABLE 6.1 List of Anticipatory Experiences—cont'd

#	Name	City	Country	Sector
74	Amsterdam Bike City	Amsterdam	NL	Sustainable mobility and transport
75	Transition Town Totnes	Totnes	UK	Trasversal sustainable approach
76	Juhnde Bio Energy Village	Juhnde	DE	**Renewable energy production**
77	Sustainable Woking	Woking Bourough	UK	Trasversal sustainable approach
78	EU ECOCITY 2002–2005	Umbertide	IT	Trasversal sustainable approach
79	Ecociudad Valdespartera	Zaragoza	ES	Sustainable housing
80	100% Renewable Energy Goal	Communauté de Communes du Mené	FR	**Renewable energy production**
81	Plan de Deplacement Urbain	La Rochelle	FR	Sustainable mobility and transport
82	Zurich's transport planning	Zurich	CH	Sustainable mobility and transport
83	Cahors – Quartier Ancien Durable	Cahors	FR	Sustainable housing
84	Ecovillaggio Torri Superiore	Torri Superiore–Ventimiglia	IT	Sustainable services and industry
85	Sunseed Desert Project	Los Molinos del Río Aguas	ES	Sustainable services and industry
86	Eco Quartier Saint Jean des Jardins	Chalon-sur-Saone	FR	Sustainable housing
87	Eco Quartier des Brichères	Auxerre	FR	Sustainable housing
88	Healing Biotope 1 Tamera	Tamera	PT	Sustainable services and industry
89	Sustainable Clonakilty	Clonakilty	IE	Trasversal sustainable approach
90	St. Davids Eco City Project	St. Davids	UK	Trasversal sustainable approach

successful[2] AEs where the production of renewable energy is the most meaningful.

Jühnde Bio Energy Village, Germany

The village of Jühnde is Germany's first bioenergy village. In 2006 they started the project with the goal to substitute all fossil fuels for electricity and heat production with biomass. The village implemented a biogas plant for combined heat and power production from liquid manure and whole plant silage of different crops. To cover the high heat demand during winter months an additional wood chip peak boiler was installed. The heat is distributed via a district heating grid providing 145 houses with heat. The electricity is completely fed into the grid.

Peccioli, Italy

The municipality launched a project that enabled environmental reclamation of the site, recycling, production of biogas, electric energy and hot water (district heating). Community stakeholding has enabled the local community to play an active role in the operations of the public company, and in the quality control of the waste disposal plant. Revenues were used to make significant investments in local services; the municipality currently produces 300% of her energy needs.

Western Harbour, Malmo, Sweden

In the Western Harbour district, polluted industrial areas have been replaced by office buildings and residential houses. The first development, Bo01, was designed to use and produce 100% locally renewable energy over the course of a year. Buildings receive energy from solar, wind and a heat pump that extracts heat from an aquifer, facilitating seasonal storage of heat and cold water in the limestone strata underground. Bo01 was the first area to use a local green space factor to promote biodiversity, incorporating local vegetation as well as rainwater through open storm water management and connection to the sea. The Western Harbour incorporates an eco-friendly transport system, with buses connecting the areas every 5 min. Bicycle lanes are easily accessible.

Samsoe, Denmark

The aim of the initiative was to base the energy production of an island entirely on renewable fuel. In just 8 years, a broad collaboration on Samsoe has managed to convert the island's energy production from oil and coal to renewable energy. Local involvement has created a bit of a social energy movement. Today, the island produces more renewable energy than it uses and exports excess energy to the mainland. Samsoe has an international lead as an energy-efficient research island, from which the Samsoe Energy Academy shares knowledge

---

2. On the basis of the assessment implemented in the frame of the MILESECURE-2050 project, high success has been achieved in 21 AEs and some success achieved in further 50 AEs.

and experiences with the world. Technologies used include: wind turbines, district heating, biomass combined heat and power (CHP) generators, solar thermal and thermal insulation of homes.

BedZED, United Kingdom

BedZED is a small district built according to environmentally sustainable criteria, impacting many areas: housing, water, waste, mobility, production of renewable energy, materials and land use. The programme's aims include living within a fair share of the earth's resources; a world in which people everywhere can lead happy, healthy lives within their fair share of the earth's resources; demonstrate what a sustainable future looks like through a variety of real-life projects; make it easy for people to understand where their environmental impacts arise and to do something about it; make one planet living be seen as a desirable and positive lifestyle choice; change the views of a critical mass of the population, including politicians, businesses and individuals, to make one planet living the norm.

6.2.4 Going Beyond Fossil Fuels in Anticipatory Experiences

One of the most important results of the study of AE in the MILESECURE-2050 project is making explicit and visible the latent role that the human factor[3] exerts in energy systems in transition. Studying the AE makes clear that, for the analysis of energy systems in transition, it is crucial to adopt a broader concept that does not just include technological aspects but also social and personal dynamics. Human energy is a holistic and all inclusive understanding, articulated in three dimensions that show different ways in which the human factor lies behind the energy system:

1. Endosomatic energy (P) represent the human capacity of effecting profound changes at the personal level in one's daily actions and convictions, in view of using the body in synergy with the energy system as a whole.
2. Social energy (S) is the human capacity to bring together different forms of social activism that coordinate and orient different social actors towards common goals and to overcome conflicts and oppositions that may represent a waste of energy.
3. Extrasomatic energy (E) is the human capacity to activate and use the natural resources through the adoption of all kinds of equipment, technology or machinery (using all energy sources, whether fossil or renewable).

---

3. In the AEs, and that represents a discontinuity and a break with the past we have observed, the rising of the human factor from an ancillary or peripheral role (which occurs only downstream in the process of change) to a lead role in the change of energy systems (upstream in the process of change) is mainly because the change of energy systems is inevitably accompanied, and at the same time is made possible only, by a deep social change.

Therefore, as we saw, if the 'weight' of renewable energy is still low in the set of the 90 AE identified, in these AE there are two other 'kinds' of energy that compete with fossil fuel that we need to take into account.

The first, the endosomatic energy, is *strictu sensu* part of energy balance and it involves directly a decrease in the use of fossil fuel. It is sufficient to consider the use of the body, first in the field of mobility (walking or cycling, a 'true' alternative of the use of fossil energy in using cars or other means of transport), but also in other fields (such as an increased use of body warmth to face the low-temperature heating system, decreasing, also in these cases, the use of other kind of energy, such as fossil fuel).

The spread of endosomatic/personal energy is very important in at least 38 of the 90 AE identified. We can quote, as a first example, the city of Bremen (Germany) where multimodal hubs bring together transit, cycling, car sharing and taxis to one location and help make Bremen a leading eco-mobile city. In the city, 60% of trips are already made by sustainable means, which comprises 14% trips by public transportation, 20% walking and 25% biking; or the city of Copenhagen (also known as 'the city of cyclists'); cycling is very widespread with nearly 400 km of cycle paths and 55% of residents in Copenhagen go to work or school by bike; or the city of Amsterdam where approximately three out of four of its residents own a bicycle, and bicycles are the most commonly used means of transport after its great efforts to promote greener means of transport.

The second, the 'social energy' is also part of energy balance but involves only indirectly decreasing in the use of fossil fuel. Social energy implies an adjustment of human and social relations that emerge in the context of the energy transition as a tendency of self-regulation. Such an adjustment allows the governance of the energy transition. Tensions and conflicts that arise in the energy systems in transition are managed through a series of continuous, coordinated and simultaneous actions, like the active participation of citizens in decision making; the widespread practice of negotiation for the resolution of conflicts and disputes between different social actors; and the ability to maintain a continuous and multilateral communication on multiple levels (from informal to institutional communication). All of these (and others similar) make energy transition effective and sustainable (otherwise it would be much more difficult). Therefore, social energy too decreases substantially the use of fossil energy.

The spread of social energy is important in almost all the AE (72 of 90), and very important in at least 35.

The analysis of the 90 AE shows that in Europe we are going and we can go far beyond fossil fuel not only because renewable energy is competing with fossil fuels, but above all thanks to the valorisation of endosomatic and social energy in an enlarged conception of human energy. This is the key (or at least one of the main keys) of the energy transition.

6.3 THE GERMAN EXPERIENCE OF THE *ENERGIEWENDE*

Germany is one of the first movers of the European energy transition and has achieved considerable progress, in particular in the electricity sector. Other areas such as energy efficiency, especially for buildings and transport, but also heating and cooling, are clearly behind and deserve more attention. The German experience is characterised by distributed renewable energy sources (RES) much more so than centralised large-scale installation.

The *Energiewende*, or energy transition, has been part of German policy discourse surrounding energy and the environment since 1980. The government policy of working towards a comprehensive energy transition was instituted in 2010, and there have been notable updates since that time, allowing for the private and public sectors to have a voice in the implementation, as well as permitting the government to adapt to changes in geopolitical realities and new energy security concerns as they arise. For example, just 3 months following the enactment of the 2010 agreement, the fallout from the Fukushima nuclear disaster in Japan renewed public support for a complete withdrawal from nuclear power. Variations of energy costs and disruptions in supply (real or threatened), particularly regarding gas from Russia, have also led to reconsiderations and reworking of the agenda.

The following sections will investigate the policy framework for the *Energiewende* and the resulting energy costs. We then move on to assess the implications for energy security using the five dimensions outlined in the book's first chapter, as well as the social implications and the impacts on acceptance.

6.3.1 Energy Policies and Support Mechanisms

The Energy Concept of 2010 created a long-term strategy for German energy policy on the path towards reducing GHG emission by 80–95% by 2050 compared to 1990 (German Federal Government, 2010). The national strategy is embedded in the EU framework on energy and climate policies, especially the EU 20-20-20 climate and energy package (Crivello et al., 2013).

The German national targets follow a hierarchy. On the top there are the political targets of minus 40% GHG emissions by 2020 compared to 1990 (minus 80–95% by 2050) and the exit from nuclear power generation until 2022 (Federal Ministry for Economic Affairs and Energy, 2015a,b,c). Below this there are the core targets of the strategy, which are:

- 40% GHG emissions by 2020; −80 to −95% by 2050 (compared to 1990)
- 18% share of RES in gross final energy consumption by 2020, growing to 60% by 2050
- 20% reduction of primary energy consumption compared to 2008 by 2020 and 50% reduction until 2050

These two main targets are further broken down into so-called steering targets. For renewable energy these are:

- 35% share of renewable energy in gross electricity consumption by 2020, growing to 80% by 2050
- 14% of renewable energy in final heat consumption by 2020
- 10% of renewable energy in final energy consumption in transport until 2020

Furthermore there are steering targets for energy consumption, which are:

- increasing energy efficiency by 2.1% per year from 2008 to 2050 as measured in energy productivity
- 10% reduction of electricity consumption compared to 2008 until 2020 and by 25% until 2050
- 10% reduction of final energy use in transport compared to 2005 by 2020 and 40% reduction until 2050

Reductions of final energy demands for space heating, measured by an increase in the refurbishment rate of the building stock, are at 2% a year (Table 6.2).

The elements of success can be summarised as follows (German Federal Ministry for Economic Affairs and Energy, 2014):

- 13.5% of gross final energy consumption coming from renewable sources in 2014;
- Renewable energy has become the dominant source for electricity in 2014 with a total generation of 161 TWh and a share of 27.4% in gross electricity consumption;

TABLE 6.2 Main Targets of the *Energiewende*

	The National Targets of the German *Energiewende*	
	2020	**2050**
Greenhouse gas emissions	−40% (compared to 1990 levels)	−80 to −95% (1990 levels)
Share of renewable energy sources (RES) in gross final energy consumption	18%	60%
Share of RES in gross electricity consumption	35%	80%
Energy efficiency	2.1% increase per year from 2008 to 2050 (in energy productivity)	

- 12% of final heat consumption coming from renewable energy;
- Avoidance of 151 million tons of CO_2 equivalent GHG emission in 2014, of which 110 million by renewable electricity;
- €19 billion investment in renewable energy installations in 2014;
- Reduction of the wholesale trading price of electricity by 10% compared to 2013 to 35.09 €/MWh in 2014;
- Avoided imports of fossil fuels worth around €9.1 billion in 2013.

The focus on generation and consumption lead to consideration of the following three areas: electricity (specifically renewable electricity generation), heat and transport. Policies implemented by the *Energiewende* seek to maintain price volatility within an acceptable boundary, a key component of a secure energy system discussed in Chapter 1. However, the German experience to date has been uneven in the ability to restructure the infrastructure and market realities to provide for both cost and supply efficiencies.

6.3.1.1 Renewable Electricity Generation

The German Renewable Energy Sources Act was adopted in 2000 and stimulated growth in renewable energy capacity and generation by establishing the following principles (Federal Ministry for Economic Affairs and Energy, 2015a,b,c):

1. Feed-in tariff: fixed and guaranteed remuneration for electricity generation from sun, wind, water or biomass for every kilowatt-hour of electricity for 20 years.
2. Guaranteed priority grid access, i.e. connection to the grid and purchase of the generated electricity are ensured.
3. Costs of RES are rolled-back to final consumers via a renewables-surcharge.

Thus, Germany is on track for its renewable electricity target, but less so for the overall renewable energy target (Fig. 6.2). This focus on electricity leads to the majority of the impacts also manifesting in this field.

6.3.1.2 Heat

The Renewable Energies Heat Act (2008 EEWärmeG) set out a target of 14% share of final energy consumption for heating by 2020 from RES (German Federal Ministry for Economic Affairs and Energy, 2014). The provision of heat energy from renewable sources amounted to 139.5 TWh in 2014 (12% of final heat consumption). However, it has remained fairly constant over the past years. The major source is from biomass (solid and other) with the fastest growth occurring in heat pumps and solar thermal collectors. Approximately 50% of new buildings use RES to some degree for securing the heat demand. It is evident, however, that the current policy is stagnating and not sufficient for a major energy transition.

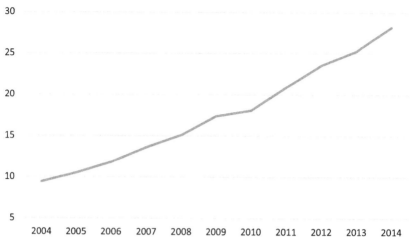

FIGURE 6.2 Electricity generated from renewable sources (% of gross electricity consumption).

6.3.1.3 Transport

Renewable energy sources face even more difficulties in the transport sector where total provision peaked in 2007 and has stagnated since then with around 5.6%, or c. 36 TWh in 2014, mostly in the form of biodiesel and bioethanol. Renewable electricity in transport is still a niche product with just 3 TWh.

With the current policy set, there is no major change expected in the transport sector and it is unclear how Germany will meet the 10% renewable energy target set out by the EU (European Commission, 2009).

The lack of development of the heat and transport sector transitions has placed more pressure on the success of the electricity sector. Although future implementations are foreseen, in order to assess the broader energy transition efforts, a more detailed accounting of the costs must be made analysing the electricity markets.

6.3.2 Electricity Costs for Retail and Wholesale

The German electricity market is split into retail and wholesale with both markets fully liberalised since 1998. Nevertheless, while wholesale customers face dynamic pricing for future and spot electricity purchases on the European Energy Exchange (EEX) in Leipzig, retail customers are usually bound by year-long fixed pricing, some with a night-time and day-time differentiation. Here again, we have to differentiate between household and commercial customers.

6.3.2.1 Retail Pricing

While the renewables surcharge is in principle to be paid by all retail customers, a number of commercial users are exempt from it (Federal Ministry for

Economic Affairs and Energy, 2014). Exemptions apply for the energy-intensive industries and the rail sector. Over 95% of manufacturing businesses are not exempt.

On the one hand the funding from the surcharge has made renewable energy one of the main sources of Germany's electricity supply but on the other hand the rapid development of renewable energy has also rapidly increased the costs paid by the consumer via the surcharge: in 2010 2 Ct/kWh; in 2015 6.24 Ct/kWh (Federal Ministry for Economic Affairs and Energy, 2015a,b,c). The total contribution to the surcharge was €23.6 billion in 2014 with households paying €8.3 billion, industry €7.4 billion and services €4.5 billion, the remainder coming from public entities, agriculture and nonrail transport (Federal Ministry for Economic Affairs and Energy, 2014). Exemptions in 2013 amounted to €4 billion, representing over 1 Ct/KWh or 19.7% of the total surcharge. Needless to say that nonexempt customers were questioning the surcharge design.

In order to better control the development of renewable energy and the associated costs, a legal reform was adopted in August 2014. The new act aims to better control the development by focusing funding on wind and solar; establishing specific deployment corridors for renewable energy; increasing the competition for the marketing of renewable sources and determining the funding by auctions (Federal Ministry for Economic Affairs and Energy, 2015a,b,c).

Thus, household electricity prices grew by 12.2% from 2012 to 2013, ie, from 26.06 Ct/KWh to 29.32 Ct/KWh, the increase was slowed considerably to 1% in 2014, ie, 29.52 Ct/KWh.

While household spending on electricity increased slightly in 2013, total spending for energy-related costs as a share of disposable income remained stable. A four-person household spent around €4070 in 2013 and €4156 in 2014, a moderate increase.

On 1 January 2015, the renewables-surcharge was reduced for the first time since its introduction to 6.17 Ct/KWh (German Federal Government, 2015), Nevertheless, household electricity prices are still the second highest in Europe, after Denmark (Eurostat, 2015a,b,c,d,e; see Fig. 6.3).

Small businesses and industry using less than 20 MWh equally pay higher than average electricity prices in Germany. Prices were 23% above the average in the second half of 2013 (German Federal Ministry for Economic Affairs and Energy, 2014). Electricity prices for nonenergy intensive medium-sized businesses and industry are also above the European average, albeit much less pronounced (Eurostat, 2015a,b,c,d,e; see Fig. 6.4).

For customers with a yearly consumption above 150 GWh, lack of data hinders a clear comparison (German Federal Ministry for Economic Affairs and Energy, 2014). Several studies estimate the costs to be very close to the wholesale average price.

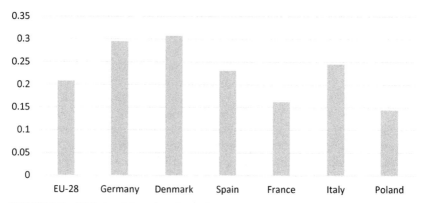

FIGURE 6.3 2015 electricity prices for medium-sized households including taxes and levies (EUR/kWh). *Eurostat.*

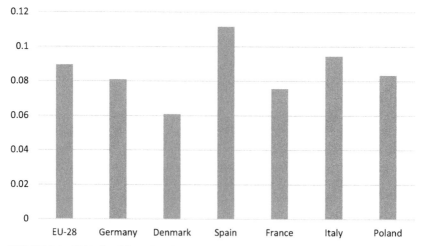

FIGURE 6.4 2015 electricity prices for medium sized businesses excluding taxes (EUR/kWh). *Eurostat.*

6.3.2.2 Wholesale Pricing

The wholesale market did not experience the same kind of price increase. On the contrary: renewable energy sources with very low marginal costs drive the wholesale prices down. From 2012 to 2013, average traded prices decreased by 20.6% for the base-load future from 49.23 €/MWh to 39.06 €/MWh and 11.7% for the spot price. In 2014 the wholesale price decreased further down to 35.09 €/MWh (base-load future).

The EEX lists just 51 entities for the electricity spot and future market (EEX, 2015). The list includes the four large German energy utilities EnBW, RWE, E.ON and Vattenfall, some municipal energy providers, some foreign energy providers, energy brokers and just a handful of industrial end-users. Even large industrial customers are buying their electricity through brokers, given the high transaction costs of getting a trading permit on the market. In practice, there is no real difference whether a company is directly participating or going through a broker, they have access to the same wholesale pricing.

Wholesale prices continue to be very low, thus providing very affordable electricity costs to wholesale customers, including also energy intensive industries.

Thus, while Germany's small-scale customers pay above-average electricity prices, low prices for energy-intensive customers contribute to a high level of competitiveness of the German manufacturing sector. According to the World Economic Forum, Germany ranked fourth in competitiveness among the world's economies in 2014 (World Economic Forum, 2015), underscoring the fact that higher energy costs and ambitious energy transition do not have to be economically detrimental.

Our main focus is, however, on the energy security implications of the German energy transition which we want to assess in the following section.

6.3.3 Energy Security Implications of the *Energiewende*

Since the beginning of Germany's energy transition, the share of gas in the final energy consumption and power generation did not increase (German Federal Ministry for Economic Affairs and Energy, 2014). The share in gross power generation was 9.7% in 2014, the lowest since 2002 (AG Energiebilanzen, 2015). Gas has a much stronger role for heating and industrial uses in Germany. Thus, given the marginal impact of the energy transition policies on heating and industrial processes, we will forego an in-depth discussion of the import dependence in terms of natural gas and resulting energy security implications but rather focus on the electricity sector as the area most affected so far by the energy transition in Germany.

Stability

Grid stability is remarkably high in Germany with 15.3 min supply disruptions reported in 2013, down from 21.6 in 2003 (Bundesnetzagentur, 2015). In 2014, the total duration of blackouts for Germany was 12.17 min, which is the shortest average interruption to supply in history. Germany's score was the third best in Europe. In comparison, disruptions were 68 min in France, 55 min in the United Kingdom, 23 min in the Netherlands and 11 min in Denmark (Federal Ministry for Economic Affairs and Energy, 2015a,b,c).

Transmission system operators had to redispatch for a total of 7965 h in 2013, 11% more time than in 2012. A total of 4390 GWh of redispatch interventions were necessary in 2013, slightly less than in 2012, amounting to less than 1% of power generation outside the feed-in tariff. Similarly, voltage-related dispatch decreased from 2012 to 2013. The reported costs were just over €130 million.

At the distribution level, four operators took 4394 h of adaptation measures over 346 days. Primary reserves cover very short term imbalances from 30 s to about 15 min. In Germany, the primary reserve amounts to about 640 MW up and down in 2014 (Stark, 2015). Demand was 551 MW in 2013, leaving significant room (Bundesnetzagentur, 2015). While stability is certainly far from perfection and disruptions persist, the problem is definitely not (yet) increasing.

On the other hand, the increased renewable energy generation puts without any doubt additional stress on the transmission and distribution system and thus endangers potentially the stability.

In the longer term, the additional envisioned grid improvements and the revision of the market design will very likely result in an overall high level of stability.

Flexibility

Secondary and tertiary reserves cover the domain between 30 s and 60 min and beyond. The available regulating capacity is about 7 GW up and 5.5 GW down in normal conditions (Stark, 2015). The secondary reserve is fully available within 5 min and is activated automatically; the tertiary reserve is activated within 15 min.

The use of minute or tertiary reserve by the transmission system operators (TSOs) decreased in 2013 to the lowest level since 2008, signalling a potential trend change and better grid stability or coordination among TSOs. The number of tertiary control dispatches decreased from over 20,000 in 2012 to about 12,500 incidents, the lowest number since 2008 (Bundesnetzagentur, 2015). In 2014, secondary and tertiary regulation decreased even further, indicating that the policies and measures reduce the need for regulating capacity (BDEW, 2015).

With the increasing integration into the European electricity grid and cooperation with neighbouring TSOs, the system flexibility will rather increase in the coming years.

Currently, however, average available cross-border transmission capacities remain low with many countries and need to be further improved. The International Grid Control Cooperation, a cooperation between Germany and its neighbours, is currently expanding the potential for integrating cross-border and inter-market electricity flows (Regelleistung.net, 2014).

Considering these trends, it is likely that grid flexibility will improve further.

Resiliency

In March 2015, a solar eclipse passed through Germany and put the electricity system under a major stress test (ENTSOE, 2015). Germany was well prepared with almost 9.7 GW available total upward positive TSO reserves (primary, secondary, tertiary and others) and almost 9 GW downwards. Within a few minutes, solar generation first decreased by 7 GW and the rebounded with 15 GW by the end of the eclipse. In the end, the grid could be kept stable within the standard frequency range.

This case study illustrates the ability of the German TSOs to adapt to transient shocks.

Robustness

In Germany the demand for electricity can be as high as 82 GW in peak hours, compared to a total installed capacity of 188 GW, including 83.3 GW in renewable energy. To ensure security of supply – especially during the high-demand winter months – the grid operators and the federal network agency have decided to create a reserve capacity: a reserve power plant capacity of up to 4 GW. Further, there is the possibility for power plants to be reactivated if security of supply is endangered. Power plant operators must notify TSOs and the federal network agency 12 months in advance of their plans to shut down a power plant (German Federal Government, 2015). On the other hand, Germany has about 10–12 GW excess capacity in the years to 2016 and can rely on 9 GW in pump storage capacity (German Federal Ministry for Economic Affairs and Energy, 2014).

The German power system is getting more diversified and available total capacity increases, as well as transmission capacity and interconnections (German Federal Ministry for Economic Affairs and Energy, 2014).

While significant additions to the transmission system are planned, current implementation lags behind. While 1876 km HV lines are planned, only 54 km have been built in 2013 and 116 km in the first three-quarters of 2014 (German Federal Ministry for Economic Affairs and Energy, 2014).

Adequacy

The liberalisation of the energy markets drastically improved consumer choice in Germany. In 2013, customers on average had a choice between 97 suppliers (BDEW, 2015). About 12% of customers switched their electricity supplier in 2013, a slight increase to 2012. The supplier change is handled within three weeks and an ombudsman is available for cases of contention. 74% of customers are satisfied with their electricity supplier, but only 40% are satisfied with the price level (BDEW, 2015).

While electricity prices remain high in Germany, functioning markets ensure consumer choice. The coming years will see a profound transition of the electricity markets with an integration of the wholesale and retail markets.

While it is safe to say that the very high number of actors in the German energy market requires more effort in coordination and cooperation, the results show the feasibility of the approach.

6.3.4 Acceptance and Social Implications

At the very latest since the Fukushima nuclear catastrophe, the approval for the *Energiewende* has been high among the population. Several surveys show the consistently high support ranging from 56% to 92% (German Federal Ministry for Economic Affairs and Energy, 2014). Ninety-two percent of Germans support the continued expansion of RES (Agentur für Erneuerbare Energien, 2014). Furthermore, 65% support renewable energy installations in their immediate neighbourhood with even higher acceptance rates for citizens who have already been exposed to local renewable energy projects. Overall, 71% believe that renewable energy protects the climate and 67% see it as a means to support energy security. Especially interesting is that 55% consider the current level of the renewables-surcharge adequate and only 36% deem it too high.

These very high support rates have multiple foundations. However, one particular aspect of the German experience is the high share of citizen-owned renewable energy investments. Over 880 energy cooperatives have invested a combined €1.2 billion and 47% of all installed renewable energy capacity was citizen-owned in 2012 (Agentur für Erneuerbare Energien, 2015a). While the share of citizen-owned energy may decrease with the rise of large-scale offshore wind projects, the contribution to the overall acceptance is beyond doubt. This can also be observed in a survey among 90 renewable energy communities which showed that acceptance is considered a key success factor in the energy transition and the implementation of local energy projects (Agentur für Erneuerbare Energien, 2015b).

The German industry association BDI runs a yearly survey among citizens and industry to assess the acceptance of the German energy transition, and it finds that just 45.3% of Germans are ready to accept negative impacts such as additional transmission cables (IW Consult, 2014). This shows that the acceptance of energy infrastructure such as transmission cables is much less pronounced. One reason for this could be linked to the lack of participation in large-scale infrastructures, ie, the lack of citizen-ownership.

The future of the German energy transition will depend on improving acceptance not just for the energy transition in general, but also for related transmission infrastructure. Alternative mitigation options are more expensive earth cables or improved design transmission line towers to reduce the visual impact.

6.4 THE ITALIAN EXPERIENCE WITH RENEWABLE ENERGY

The Italian experience with RES is strongly related to the recent evolution of the photovoltaic (PV) sector and of solar energy, more generally. The PV sector, in particular, experienced a heavy boom in terms of installed capacity in the last years, mainly because of the generous incentive schemes implemented at a national level. This certainly led to an increasing contribution of solar energy to satisfy energy demand, but had the drawback of rising concerns in terms of technology dependency from abroad (mainly China and Germany) since the domestic industry did not prove ready to compete in the PV-module segment of the value chain.

Since 2000, the share of RES in the national electricity generation mix increased from 18% to almost 39% with a substantially stable trend in the period 2000–2008 and a pronounced upward trend in the period 2008–2013 (Fig. 6.5).

In the period 2011–2013, the contribution of fossil fuels, notably of natural gas, to the national electricity generation mix decreased substantially. The drop in the share of natural gas was more than compensated by the rise in the share of RES that in 2013 overtook natural gas as the main energy source for electricity generation in Italy.

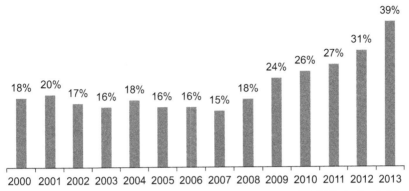

FIGURE 6.5   Share of RES in the Italian electricity generation mix, 2000–2013. *Authors' elaboration based on GSE (2015) data.*

Historically, the trend of electricity generation from RES has been mainly connected to the contribution of large-scale hydro power plants. As Fig. 6.6 suggests, however, in the period 2000–2013, electricity generation from RES more than doubled mainly because of the prominent role gained by the 'new' RES, in particular solar energy.

At the end of 2013, a total of approximately 600,000 renewable energy plants were installed in the Italian territory (Fig. 6.6): 99% of these plants are solar PV plants, most of which (58%) are small-scale plants with a nominal power between 3 and 20 kW. Solar PV plants account for approximately 36% of overall RES installed capacity (more than 18 GW out of almost 50 GW) and electricity generation from solar covers approximately one-fourth of the overall electricity generation from RES. Territorial differences are important in Italy in terms of both electricity generation and network infrastructure. In 2013, the Puglia region (in the South of Italy) was the main contributor to electricity generation from PV plants (approximately 17% of the national value) whereas most of the electricity demand originated from the North of the country.

6.4.1 Energy Policies and Support Mechanisms

On 8 March 2013, the Italian Ministry of Economic Development jointly with the Ministry of the Environment approved the document containing the new National Energy Strategy (Strategia Energetica Nazionale, SEN). The final version of the document is the result of a public consultation that started in October 2012 and that involved, directly and indirectly, a wide range of stakeholders (institutions, research centres, associations and nongovernmental organisations [NGOs], social partners, industry and economic actors of the energy sector). It outlines the most relevant strategic objectives for the national energy sector from a medium (2020) to long-term (2050) perspective. The actions proposed aim at defining a low-carbon growth path for the national economy while

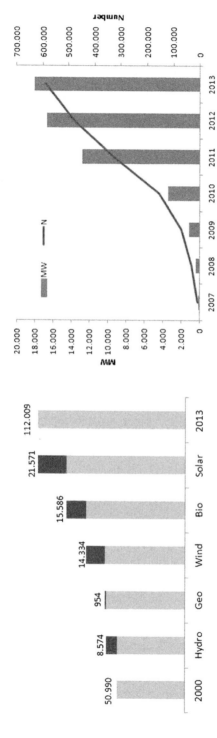

FIGURE 6.6 Evolution of electricity generation from RES (left, in GWh) and of PV plants (right, number and installed capacity in MW) in Italy. *Authors' elaboration based on GSE (2015) data.*

reinforcing the country's security of supply. In particular, energy efficiency, the sustainable development of RES and the deployment of smart grids for electricity distribution are among the main priorities listed in the document.[4]

In order to achieve the targets on GHG emission reduction and decarbonisation of the energy sector defined at the EU level, Italy has adopted a series of measures and policies at different levels and covering different sectors with the aim of: (1) completing the liberalisation process of the electricity and gas markets; (2) promoting energy efficiency; and (3) fully deploying the potential of RES to diversify the national energy mix.

In particular, the National Action Plan for the deployment of RES (PAN), issued in June 2010, implements the Directive 2009/28/CE, that established a common framework to promote RES and fixed binding national targets for the overall share of renewable energy over gross final energy consumption: for Italy, the target amounts to 17% by 2020. The PAN defines sector targets and the measures aimed at achieving them, and takes into account the possible interactions and synergies with energy efficiency measures. More specifically, the PAN defines the following targets to be satisfied by RES by 2020:

- 10% of the energy consumption in the transport sector, thanks to higher contribution of biofuels and to interventions aimed at deploying the electric vehicles sector;
- 26% of the energy consumption in the electricity sector, by means of energy storage systems, improvements in the distribution networks and the realisation of smart grids;
- 17% of air-conditioning consumption, through actions aimed at deploying district heating and cogeneration and at inserting biogas in the natural gas distribution network.

In order to incentivise electricity generation through RES, a number of measures are currently in place in Italy:

- The 'green certificates' (Certificati Verdi) system, based on a compulsory share of additional electricity generation by means of renewable energy technologies.
- Feed-in tariffs for the electricity produced and put into the grid by RES plants with a capacity below 1 MW (200 kW for wind), as an alternative to the green certificates.
- Dedicated incentives to PV and solar thermodynamic plants through the 'Conto Energia' mechanism.
- Simplified selling procedures for electricity produced by renewable energy plants ('ritiro dedicato').

The most relevant measure that helps in explaining the rapid increase in PV installations is the feed-in scheme 'Conto Energia', the programme which

---

4. For additional information and the full text of the document, see www.sviluppoeconomico.gov.it/.

grants incentives for electricity generated by PV plants connected to the grid. Italy introduced this support scheme in 2005 (Ministerial Decree of 28 July 2005 – first feed-in scheme) and the scheme is now regulated by the Ministerial Decree of 5 May 2011 – fourth feed-in scheme, applying to plants commissioned between 1 June 2011 and 31 December 2016. Under the scheme, PV plants with a minimum capacity of 1 kW and connected to the grid may benefit from a feed-in tariff, which is based on the electricity produced. The tariff differs depending on the capacity and type of plant and is granted over a period of 20 years. For plants commissioned before 31 December 2012, the scheme (called 'feed-in premium') provides for a tariff for the electricity produced. The electricity fed into the grid may be purchased by GSE ('ritiro dedicato') or economically offset with the value of electricity withdrawn from the grid (net metering, 'scambio sul posto') service. Starting from the first half of 2013, the tariff is made up of both the incentives and the value of electricity. A specific tariff is applied to the self-consumed electricity. Different parties may apply for the feed-in scheme: individuals, organisations, public entities, noncommercial entities, owners of single or multiple housing units.

6.4.2 Energy Costs for Retail and Wholesale

In the last years the deep penetration of RES in the national electricity generation mix had repercussions on the energy costs for domestic and industrial users.

Eurostat data for the second semester of 2014 show that gross (ie, all taxes and levies included) electricity prices for Italian households are on average 14% higher than the EU-28 levels. More specifically, as Fig. 6.7 suggests, Italian electricity prices for domestic users are relatively lower than the EU-28 average for small consumers (Bands DA and DB, consuming less than 1000 kWh and between 1000 kWh and 2500 kWh, respectively) whereas they are substantially higher for large consumers (the differential between Italian and EU-28 prices rising to approximately 43% for Band DE, ie, for electricity consumption exceeding 15,000 kWh per year).

In the case of industrial consumers, Italian gross electricity prices are on average 24% higher than the EU-28 levels. Such a differential is more or less stable across the different bands of consumption, with the exception of very large

	Band DA	Band DB	Band DC	Band DD	Band DE
European Union (28 countries)	0,3186	0,2245	0,2079	0,1994	0,1903
Germany	0,4336	0,3230	0,2974	0,2827	0,2714
France	0,2906	0,1955	0,1751	0,1615	0,1564
Italy	0,2913	0,2099	0,2338	0,2935	0,3337
United Kingdom	0,2503	0,2230	0,2013	0,1817	0,1679
Italy-EU28 differential	**-9,37%**	**-6,96%**	**11,08%**	**32,06%**	**42,97%**

FIGURE 6.7 Electricity prices for domestic consumers: Italy–EU-28 differential. *Eurostat, 2014; Authors' elaboration based on Eurostat (2015f) data.*

industrial consumers (Band IF: 70,000 MWh < Consumption < 150,000 MWh) where the Italy–EU-28 differential reduces to 15%.

The present situation depicts the eventual results of the recent changes in the electricity market resulting from the rapidly increasing share of RES in the national electricity mix. Hence, the time evolution of electricity prices for domestic and industrial users can help in shedding light on the effects of RES in the national market.

As Fig. 6.8 suggests, the trend of the Italy–EU-28 differential for the electricity prices of domestic consumers is more or less stable for the largest consumption band (Band DE, in yellow) whereas the differential is decreasing over time for the intermediate and, more heavily, for the lowest consumption bands (DC in red and DA in blue, respectively). In the case of the band DA, the Italian electricity prices exceeded by almost 10% the average EU-28 price in 2008, whereas in 2014 they are almost 10% lower. For the band DC, which accounts for electricity consumption in line with that of an average household, the differential decreased from above 20% to almost 11%.

If we look at the time evolution of the difference between the electricity prices of large versus small domestic consumers (Band DE-DA differential, Fig. 6.9), in the case of Italy (blue line) the upward trend indicates that, starting from a situation where the price advantage accrued to large consumers, small consumers now pay a relatively lower price for electricity than large consumers. This trend is not consistent with the EU-28 situation where the price advantage is still largely in the hands of big consumers. This might be one of the repercussions of the increasing competition of PV plants and self-electricity production for electricity supply to domestic users.

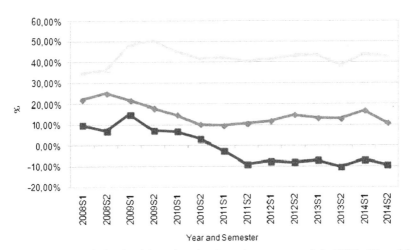

FIGURE 6.8 Trends in electricity prices for domestic consumers: Italy–EU28 differential. *Authors' elaboration based on Eurostat (2015f) data.*

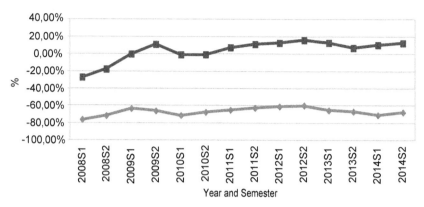

FIGURE 6.9 Trends in electricity prices for domestic consumers: Italy–EU28 Band DE-DA differential. *Eurostat, 2008–2014.*

In the case of industrial consumers, electricity prices show a more relevant decrease for large consumers than for small ones with the relative gap (Band IF-IA differential) increasing from a value of almost 125% in 2008 to approximately 150% in (Eurostat, 2015g). In this case, the market power on the demand side of big industrial consumers might be an explanation of this trend that does not seem to be related to the penetration of RES in the national electricity mix.

One of the preferred arguments of the opponents to incentivising RES in Italy is that incentives have a strong incidence on the price paid by domestic consumers. It is thus informative to decompose the price of electricity into its different components. As for the second quarter of 2015, the reference price of electricity for an average domestic consumer (household with consumption equal to 2.700 kWh/year and nominal power at 3 kW) amounts to 18.43 €/kWh (source: AEEG[5]) that can be decomposed as follows:

- 43.83% of the overall price for sale services (price of the electricity delivered to the final customer);
- 17.99% for network services (tariffs for transportation, distribution and measurement of electricity);
- 13.47% for national taxes;
- 24.71% for system duties ('oneri di sistema').

System duties include duties for nuclear security and territorial compensations, energy efficiency promotion, special tariffs for railway sector, compensations for small electricity producers, support for electric system research, electric bonus, contribution to electricity-intensive industries, as well as the incentives to RES (the so-called A3 component) that account for slightly more than 80% of the overall system duties.

---

5. www.autorita.energia.it/it/consumatori/bollettatrasp_ele.htm#dettaglio.

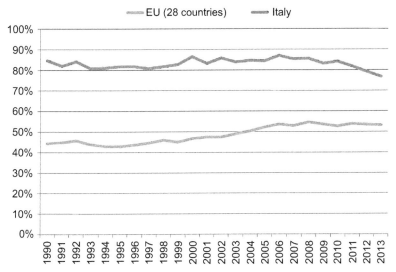

FIGURE 6.10 Trends in energy dependency: Italy versus EU-28. *Eurostat.*

6.4.3 Energy Security Implications

The strong and rapid penetration of intermittent RES in the national electricity generation mix raises several issues related to the resilience and technical capacity of the system to accommodate supply from decentralised plants and properly satisfy the demand.

One important point to note, however, is that Italy is becoming less energy dependent. Fig. 6.10 shows the trends in energy dependency for Italy and for the EU-28 countries over the period 1990–2013[6]: while Italy still displays very high levels of dependency on imports of energy from abroad compared to the EU-28 average (in 2013, 77% versus 53%), in the last years its energy dependency is decreasing in contrast to the EU-28 trend that shows a quite stable increase since the end of the 1990s.

The rising contribution of RES for electricity generation also has repercussions on the national electricity market in terms of prices. In 2014, the national unique price of electricity ('prezzo unico nazionale', PUN) reached its minimum level since the start of the liberalised market in Italy, with a value of almost 52 €/MWh. The reasons for this situation are related to at least three aspects:

---

6. Energy dependency shows the extent to which an economy relies upon imports in order to meet its energy needs. The indicator is calculated as net imports divided by the sum of gross inland energy consumption plus bunkers. Source: Eurostat, http://ec.europa.eu/eurostat/tgm/table.do?tab=table&plugin=1&language=en&pcode=tsdcc310.

1. Decrease in the cost for electricity generation from natural gas, which represents the reference fuel for the national generation park and whose cost is decreasing.
2. Shrinking of electricity demand: the slow progress in economic activity after the 2008 crisis together with important energy efficiency policies led to a decrease in electricity demand in Italy in the last years.
3. Electricity generation from RES, notably from solar and wind, implies a decrease in the average generation cost, since these sources have marginal generation costs that are close to zero.

In such an overcapacity situation, due to the strong penetration of RES, notably of solar PV, the ratio between peak and off-peak prices during working days substantially decreased from a value of approximately 2.0 in 2005 to almost 1.2 in 2014, reaching one of the lowest values in Europe. This result could be observed across all the six electricity market zones (North, Centre North, Centre, South, Sicily and Sardinia), but it was more evident for the Southern and island zones, where the share of electricity generation from intermittent RES is larger.

Another important effect of the strong penetration of RES in the electricity generation mix is represented by the price volatility dynamics, differing across zones, and strongly increasing mainly in the Southern regions where RES power plants are highly concentrated (Fig. 6.11).

The rapid and deep change of the Italian electricity generation system in the last 10 years, beside its impact on the electricity market, also raised a series of issues related to the normal and proper functioning of the system itself: first of all, the system capacity to balance demand and supply curves, and to do it promptly and effectively. The robustness of the system needs indeed to be tested with respect to its capacity to respond to quick and relevant unbalances and to properly manage the coexistence of electricity generation from intermittent RES with that from traditional fossil fuels.

Electricity generation from RES is strongly supported on the economic side through different measures but also through some rules governing the functioning of the electricity market, such as the dispatch priority for the electricity produced by RES power plants. As a result of the increase in RES electricity, the number of functioning hours of traditional fossil fuel plants decreased significantly. This eventually led to the closure of some of them and to a revision of the strategic choices of many operators and required the rest of the system to have versatility and flexibility to guarantee the prompt satisfaction of energy demand.

In this sense, the national electricity system needs to have:

- The capacity to face most of the electricity demand in the hours where RES cannot produce electricity (eg, night hours for PV plants) through nonintermittent power plants. Since one of the two demand peaks is at 7 pm, this implies an important effort for the electricity system in terms of response through controlled-production plants or stock systems (peak load adequacy).

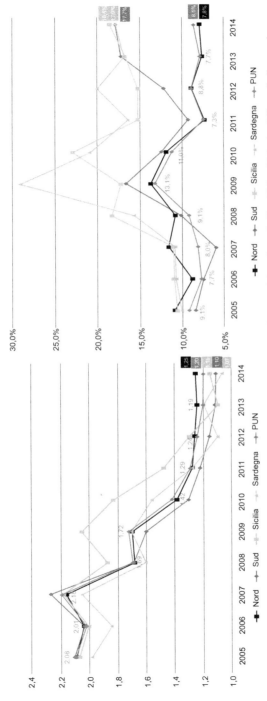

FIGURE 6.11 Evolution of the ratio between peak and off-peak electricity price (left) and of the price volatility (right, %) in the different Italian market zones. *GME (2014).*

- The capacity to balance the load at the minimum (minimum load balancing), ie, to minimise the generation based on fossil fuels during the functioning hours of RES plants in order not to have an overproduction (negative residual load).

The proper functioning of an electricity generation system characterised by a strong presence of intermittent RES plants, such as the Italian one, is thus affected by:

- The flexibility of traditional power plants to rapidly enter into operation in case of missing generation from RES plants or of rising demand or both simultaneously: this requires very high 'rampa' rates, ie, the speed at which a number of plants can enter into operation, estimated in the order of 10 GW/h (up to 40 in the absence of wind energy).
- The predictability of production: while the uncertainty about electricity demand is typically in the range of 1–2% of the overall load, the average absolute error for wind productivity is around 15%, 24 h before in real time.

Fig. 6.12 shows the change in the way the electricity demand was satisfied in Italy (South zone) in 2010 and in 2013. In a 3-year period, the demand curve moved down and the share of demand covered by intermittent RES (in yellow) increased substantially, thereby leading to:

- The risk of overproduction during the solar radiation hours: this happened in the following years both in working and in nonworking days.
- The need for several traditional plants to enter into operation very rapidly at some point in the day because of the simultaneous increase in demand and decrease in production from PV plants.

6.4.4 Acceptance and Social Implications

The growing contribution of RES to electricity generation puts pressure on the ability to accommodate into the grid decentralised and intermittent power generation plants with respect to traditional fossil-fuelled centralised power plants. In just 1 year, between 2010 and 2011, the number of small-scale power plants for electricity generation more than doubled (from 159,878 to 335,318). Almost 80% of the so-called distributed generation (DG) originates from RES, notably from PV plants, and accounts for approximately 10% of the overall national electricity generation. In particular, in 2011, gross electricity generation from DG power plants reached 29.2 TWh, rising by 47.4% with respect to the previous year, mainly thanks to the contribution of PV plants (+458.3%). Such a rapid development of small-scale power plants fuelled by intermittent RES and connected to the grid calls for a likewise rapid evolution of regulation in order to integrate in a sustainable and secure way these plants into the electric system. In this sense, the National Authority for Electricity and Gas points out that, on the one hand, network and grid infrastructures need to be properly redesigned, and,

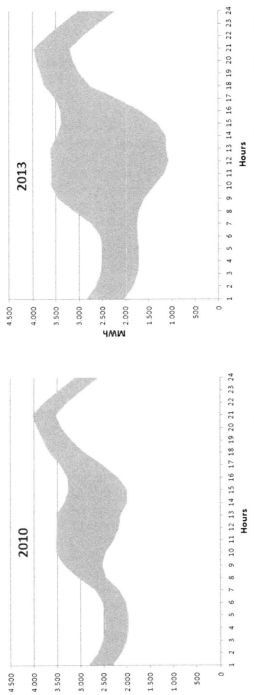

FIGURE 6.12 Daily electricity supply by intermittent RES (in yellow (grey in print versions)) in Italy (South zone) in a standard working day, years 2010 and 2013 (in MWh). *Authors' elaboration based on GME (2014) data.*

on the other hand, innovative ways for managing the network need to be agreed upon with all the stakeholders, including of course private citizens.[7]

Bottlenecks arise in the grid also due to the fact that RES electricity generation is mostly concentrated in the Southern Italian regions and in the Islands whereas most of the electricity demand comes from Northern regions where most of the industrial production takes place. In order to modernise a grid that in recent years has also experienced episodes of severe disruption of energy provision, Terna, the Italian grid operator, began a series of investments aimed at realising 2500 km of new power lines and 84 new stations for strengthening national and international connections. As for the Northern interconnections, new lines are going to be realised with France, Austria and Slovenia. In the South of Italy, the power line Sorgente-Rizziconi, scheduled to improve internal connection between Sicily and Calabria, could become a pillar to realise an Italian 'electricity hub' of the Mediterranean. The political events and the history of the country in recent years have fuelled the growth of an antipopular sentiment against the so-called major projects. In this sense the power line Sorgente-Rizziconi represents an example of the dangerous liaisons between a project, the territory and its inhabitants. Administrative procedures including environmental impact assessment have been completed in accordance with the conclusion of the work planned by the year 2015. In March 2015, the judiciary authority has intervened with the seizure of one of the pylons as a result of failure to comply with the legislation on landscape restrictions and the distance from residential areas. This act is a result of the protests carried out for years by environmental groups and local committees engaged in defending citizens' health and the territory. According to Terna, this stop could generate considerable economic losses and social damage due to the risk of blackout for aged power lines. In fact, the acceptance of energy infrastructure is key for the success of the energy transition process that Italy is undergoing. But this is something that is still far to be reached.

In spite of this strong and widespread sentiment against big infrastructural projects, the extraordinary boom of RES in Italy led to a growing awareness of the role that energy consumers can have also as producers of energy, ie, of the new figure of 'prosumers'. As pointed out above, the Italian territory recently experienced a growth in the generation of clean and distributed energy, realised with more efficient and integrated systems of self-production or production and distribution at local level (Rapporto Comuni Rinnovabili 2015, Legambiente). Although the Italian experience with the so-called energy cooperatives is not as widespread as in Germany, there are several examples that show the rising interest in alternative forms of financial cooperation for investing in RES. In most cases, groups of citizens from the same municipality get together and buy PV plants located in their territory. The Energyland Cooperative in Verona accounts

---

7. 'Settimo Monitoraggio della Generazione Distribuita', 2013, Autorità per l'Energia Elettrica e il Gas.

for almost 100 households and since the end of July 2011, it was able to generate more than 3 million KWh of renewable energy. The Local Solar Community (Comunità Solare Locale) is an association of citizens and firms born in 2011, thanks to the pilot project Sistema Integrato di Gestione dell'Energia (Integrated System for Energy Management; SIGE), promoted by the University of Bologna and co-financed by the Emilia-Romagna region. Solar Communities represent groups of households and citizens that share with their own municipalities and local authorities the objective to decrease energy consumption and GHG emissions while increasing the use of energy generated by RES. Besides being a way to involve and increase the awareness of citizens towards a more sustainable use of energy, Solar Communities are based on alternative mechanisms of financial participation to energy generation projects. These mechanisms allow citizens to produce the share of renewable energy that they need for their domestic uses without incurring high investment costs. The Sun in Network (Sole in Rete) project is a purchasing group for PV plants, that counts almost 6000 members, 1852 plants and 8760 kWp installed capacity. Another relevant example, that is also one of the AEs investigated in the MILESECURE-2050 project, is represented by the municipality of Peccioli in Tuscany that, with the project Un Ettaro di Cielo, realised a 1000 kWp PV plant with the joint participation of the municipality and private citizens as local stakeholders and shareholders.

The public co-financing of these initiatives, together with the low capital involvement required to private citizens, may be key factors for the potential and successful replication of these experiences in Italy and across Europe, also as a way to deal with the recent and rising phenomenon of energy poverty.

6.5 COSTS, COMPETITIVENESS AND CLIMATE CHANGE MITIGATION IN EUROPEAN UNION ENERGY SECURITY POLICY

As was pointed out in Chapter 1, the EU has been unable to implement a comprehensive energy security policy, due particularly to its inability to override national energy policies of the Member States and practically to uncertainty in the relevant working definitions of energy security. The effect has been the merging of disparate national strategies into a comprehensible narrative that is often reactive rather than proactive. Specifically, these national strategies find themselves responding to small- or medium-scale temporal issues or price fluctuations, rather than addressing the holistic objectives an EU-wide policy would be able to address. Moreover, the uncertainty regarding systems integration, whether infrastructure interconnectivity or energy markets, leads to further delay in action. These delays propagate further insecurity for Europe's energy future.

In this section, we will address the lessons learned from a decade's worth of EU energy transition efforts, highlighting how climate change mitigation strategies, energy costs and market-oriented policies will create a new set of implications for EU energy security. By considering the five-point framework established in the first chapter and applied throughout this book, we will seek to determine the ways in which EU climate policies already implemented and their measurable impacts are serving to transition the EU energy system, in as much as it can be considered singularly, towards a secure future.

6.5.1 European Energy Transition

The European Commission notes that Europe is in need of an energy transition for several, but especially the following, reasons (European Commission, 2015g):

- The EU imports 53% of its energy at a cost of €400 billion.
- Six Member States import their natural gas from one single exporter, making them extremely vulnerable.
- 94% of transport is oil-dependent, of which 90% is imported, making transport the least energy-independent sector.

On the other hand, the European Commission is far from having the authority to set the national energy policies for each of its Member States and, instead, must focus on coordinating and integrating the highly disparate European energy systems. As discussed more in detail in Chapter 2, in order to achieve progressive measurable change, the EU formulated a number of policies and targets.

Going beyond the 20-20-20 targets, the new 2030 energy climate strategy aims for a reduction of GHG emissions by 40% by the year 2030 compared to 1990 (European Commission, 2014a). In addition, the strategy calls for a minimum of 27% renewable energy consumption, 27% energy savings compared to the business-as-usual case, electricity interconnection between Member States of at least 15% of installed capacity and a reliable climate and energy governance system.

The Energy Union, the framework for which was laid forth in 2015 and seeks to bring Europe's energy policies and infrastructure in line with the overarching security, climate and efficiency agendas (European Commission, 2015g), calls for four additional energy priorities:

- Energy security, solidarity and trust;
- A fully integrated European energy market;
- Research, innovation and competitiveness; and
- Energy governance reform.

Apart from the climate and energy targets, all other Energy Union objectives have a far softer formulation, making the tracking of progress more cumbersome.

On the climate target, the EU is currently on track with emissions in 2014 23% below 1990 and 4% under 2013 levels (European Commission, 2015h). According to their own projections, emissions will be 24% below 1990 levels in 2020. However, for 2030, current policies only achieve 27% reduction, ie, 13% are still missing and require actions both via a reform of the EU Emissions Trading Scheme (EU ETS) and new initiatives.

The reform in the EU ETS aims for total emissions reductions in the covered sectors of 43% by 2030 compared to 2005, to be achieved by annual reductions of the available allowances of 2.2% from 2021 onwards, a steeper reduction compared to the current yearly target of 1.74%.

The EU supports climate mitigation by focusing 20% of its budget on related actions, corresponding to €180 billion from 2014 to 2020, including over €110 billion in the European Structural and Investment Funds (European Commission, 2015f).

Progress in the deployment of renewable energies is mixed (European Commission, 2015c): in 2014, 15.4% of gross final energy consumption came from renewable energies, basically on track towards the 2030 target; 26% of electricity came from renewable sources, including 10% from variable sources such as wind and solar. In the transport sector, progress is particularly slow with currently only 5.7% renewable energy compared to the 10% target for 2020. In heating, the share is at an encouraging 16.6%, but is still expected to be short of the 2020 target.

So far, only one Member State, Sweden, has a cross-border renewable energy support mechanism (European Commission, 2015f). Installed renewable electricity capacity in the EU reached 380 GW in 2013, compared to 450 GW in fossil fuel plants. In 2014, 12.4 GW of new wind capacity were added, a new record high (European Commission, 2015c).

With regard to energy efficiency, the EU is projected to reach just 17.6% primary energy savings compared to projections for 2020, well short of the pathway towards the 27% target (European Commission, 2015f). It seems that the majority of climate change mitigation efforts will come from additional renewable energy generation, rather than from energy efficiency improvements.

The future additional increase of renewable electricity will require additional transmission infrastructure. To help integrate the European electricity market and improve the interconnection between Member States, the European Commission collected a list of 248 projects of common interest (PCIs), both for electricity and natural gas networks and will benefit from €5.35 billion support via the Connecting Europe Facility, leveraging additional public and private funding (European Commission, 2013).

In 2014, 34 projects were selected with €647 million funding; in 2015, €149 million in funding have gone to 20 projects so far. However, the PCIs lag clearly behind schedule: 13 projects are expected to be completed by the end of 2015 and more than 100 are in the permitting phase, over 25% are delayed (European Commission, 2015f).

The following interconnections have been achieved in the first year of the Energy Union (European Commission, 2015f):

- electricity cable between Italy and Malta in April 2015, ending the energy isolation of the island
- completion of Eastlink between Finland and Estonia and of Nordbalt between Lithuania and Sweden, both enabling the integration of the Baltic States into the NordPool energy market
- a new France–Spain electricity interconnector, doubling the transmission capacity between the two countries

In the gas market, a key concern of security of supply for Europe, the following progress has been made:

- a new liquefied natural gas (LNG) terminal in Klaipeda (Lithuania)
- gas interconnector between Hungary and Slovakia
- important reverse flow equipment was installed within the EU as well as on its borders to Ukraine facilitating bidirectional trade

Nevertheless, 40% of EU gas imports came from Russia in 2013 with several Member States still fully or almost fully dependent on supply from Russia, especially Bulgaria, Czech Republic, Estonia, Finland, Hungary, Latvia, Lithuania and Slovakia.

Taken together, the European energy transition has made considerable progress already, but has also significant challenges to yet overcome. Some of these barriers require deeper restructuring of the energy market design, both for electricity and for gas, with different objectives in mind.

The reshaping of energy governance will be the key cliff for the EU to take, not the seemingly difficult infrastructure projects which, in the end, only require money. Should the EU be positioned to administer and implement the governance changes necessary, it is increasingly likely that individual Member States and private investors will provide the means for ensuring project completion. In this sense, security of governance and certainty of policy will go a long way to providing security for the future for an integrated European energy system.

6.5.2 Energy Costs for Retail and Wholesale

Public and private sector buy-in of the energy transition hinge, though not exclusively, predominately on costs. In order to ensure support for a broader energy transition, energy costs, both at the retail and wholesale levels, must stay within the qualitative boundaries of acceptable variance, as discussed in Chapter 1. Here again we see how the term 'security' denotes several perspectives. Market integration is essential to maintaining security for consumers on the EU level; the process for getting to there is much more complicated. Additionally, volatile energy prices on the wholesale market compromise competitiveness. In order to evaluate the boundaries and develop integration policies that do not jeopardise this security, the European Commission reviewed price trends between 2008 and 2012 in their 2014 report, focusing on electricity and gas (European Commission, 2014b).

6.5.2.1 Electricity

Household electricity prices have risen by 4% per year over 4 years from 2008 to 2012, with significant differences between the Member States. Moreover, household electricity prices differ significantly from one country to another with Denmark at the helm with 30 Ct/KWh, Bulgaria at less than 10 Ct/KWh and the EU average at 20.8 Ct/KWh in 2015 (Eurostat, 2015a).

Over the 2008–2012 time horizon, industry electricity retail prices rose slower at 3.5% per year on average. Prices in 2015 were highest in Malta with 15.5 Ct/KWh for medium sized-companies and lowest in Sweden with 6.2 Ct/KWh, the EU average being at 8.9 Ct/KWh.

Wholesale prices for electricity converged and dropped between 35% and 45% in Europe between 2008 and 2012. Reasons for this trend are three-fold: (1) increased coupling of markets; (2) unbundling of electricity generation from system operation; (3) the fall of EU ETS carbon prices; and (4) the growth of renewable power generation with low operating costs.

The lack of connection between wholesale and retail markets points to inefficient markets.

6.5.2.2 Natural Gas

Natural gas prices for households rose on average 3% per year between 2008 and 2012. In 2015, gas prices ranged 31.43 €/GJ in Sweden to as low as 8.65 €/GJ in Romania, with the EU average at 18.44 €/GJ (Eurostat, 2015b).

For industry, gas prices rose 1% per year on average from 2008 to 2012. In 2015, the highest gas price for medium-sized industry was 14.1 €/GJ in Liechtenstein, the lowest 5.75 €/GJ in Romania with the EU average at 9.25 €/GJ. Gas wholesale prices saw both a fall and a rebound over the 2008–2012 timeframe, not resulting in any clear trend (Figs 6.13 and 6.14).

6.5.2.3 Overall Energy Costs

Both gas and electricity price developments vary considerably between Member States and between sectors. The resulting price signal is composed to some degree by the energy and supply costs and increasing by network costs and taxes and levies. Thus, European energy prices become less and less dependent on the resource input costs and increasingly dependent on infrastructure costs, taxes and levies.

For European households, this translates, despite declining energy consumption, into higher expenditures for energy: the share of consumption spending on energy rose from 5.6% to 6.4% over the 2008 to 2012 timeframe (Fig. 6.15). Since low-income households spend a higher share of their income on energy, the price increase is especially felt by the most vulnerable.

For European industry, electricity costs rose by 4%, despite a reduction of consumption by 4% from 2008 to 2012. On the other hand, total gas costs declined by 6.8%, more than the consumption decline of 5.3%.

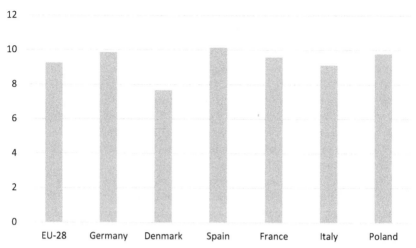

FIGURE 6.13 2015 natural gas prices for medium-sized businesses excluding taxes (EUR/GJ). *Eurostat.*

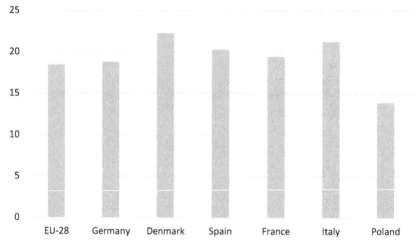

FIGURE 6.14 2015 natural gas prices for medium-sized households including taxes and levies (EUR/GJ). *Eurostat.*

While natural gas and retail electricity prices for industry are considerably higher than for many competing economies, European electricity wholesale prices are comparable to, eg, the United States. On the other hand, electricity supply is more reliable in the majority of Member States than in the United States, Russia or China.

6.5.2.4 Energy Security Implications

The energy transition strategy of the EU has been governed primarily by efforts to manage the costs of climate change mitigation policies, however

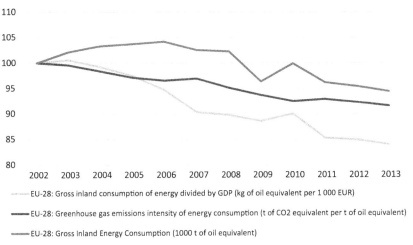

FIGURE 6.15   European energy consumption, energy intensity and GHG intensity. *Eurostat.*

possible that is with the varying national energy policies of the Member States. The Energy Union is poised to change this, placing the EU in a position to facilitate system integration to enhance interconnectivity, reduce imports and provide security. When determining the steps forward for such integration, it is valuable to assess the impacts on energy security currently implemented actions have had.

Applying the following five-point framework, we are able to see how energy security is enhanced through price balancing and climate mitigation strategies. Further integration, both of infrastructure and markets, will continue to improve the situation. However, as stated above, it will be the reshaping of energy governance, a resolute path towards interdependence within the EU that will lead to low-carbon energy security.

Stability

Short term instability is mainly an issue pertaining to the electricity grid. Total average interruptions in Europe range considerably between 15 min and over 400 min per year (CEER, 2015), exemplifying the very diverging quality of supply across the Member States. Over the 2009 to 2013 timeframe, 12 Member States had average disruptions in excess of 200 min per year.

This data points towards considerable need for added system stability. The European energy transition is improving the grid stability with added interconnectors and improved cross-border coordination among the transmission system operators.

Flexibility

In the electricity sector, the number of interruptions is decreasing in most Member States to below four in 2013 (CEER, 2015). However, the vast majority of interruptions link to low and medium voltage systems, indicating the need to improve local distribution services. At the national level, improved interconnection and coordination has significantly improved system flexibility.

In the gas sector, improved interconnection and storage are essential contributors to flexibility. However, storage is concentrated in a few Members States – Germany, Italy and France – with only limited added storage planned outside the United Kingdom (European Commission, 2014d).

Resiliency

The growing share of renewable electricity generation considerably diversifies the power generation and improves system resiliency considerably by increasing the number of power plants and their geographical distribution (European Commission, 2014c).

In the gas domain, this development has yet to occur. The implementation of reverse flow infrastructure is a crucial step towards improving system resilience. Moreover, LNG terminals can provide additional resilience by offering alternative sourcing. Past estimates show that a disruption of the gas supply in winter time would affect the majority of Member States (European Commission, 2014d). The recent addition of LNG capacity as well as improved pipeline interconnectors has certainly improved the resiliency.

Adequacy

Adequacy of the electricity market has greatly improved recently with the further integration of European power trading. Further cooperation between TSOs has reduced cross-border loop flows resulting from uncoordinated balancing. It can be expected that the situation will further improve. The European network of transmission system operators for electricity (ENTSO-E) expects that for the years up to 2025, adequacy is considered high, with just four to five Member States structurally importing electricity to ensure adequate spare capacity. However, the number of countries of Member States depending on cross-border imports to maintain adequacy is expected to increase over time from 12 in 2016 to 16 in 2015.

In the gas domain, storage capacity is still insufficient. Total LNG capacity remains low and thus adequacy depends mostly on available spare pipeline capacity (European Commission, 2014d).

Robustness

Actors in the energy market are allowed to choose from primary energy sources at cost-oriented prices, without being hindered in their choice by economic or (geo)political constraints or enduring pressures on energy sources and infrastructures over a years-to-decades horizon.

For electricity, market design differs by Member State with two Member States having only one generating entity and many countries having concentration rates beyond 60% (Eurostat, 2015c). Despite market coupling in north-west Europe, there remain significant price differentials between Member States: average base-load prices ranged from 22 €/MWh in Sweden to 58 €/MWh in the United Kingdom in the second quarter of 2015 (European Commission, 2015a). Overall, NordPool prices were by far the lowest, while Central Western Europe and Central Eastern Europe were equally below the average price as opposed to higher Spanish, Italian and Greek pricing. It can be concluded that the market coupling is far from complete and that limited market integration and physical interconnection still hinder the free flow of electricity.

For gas, the European integrated market is slowly emerging. Wholesale gas prices converged in 2015 with still significantly higher prices for Russian gas to Lithuania and Algerian gas to Italy, but also for Estonia, Romania and Slovenia (European Commission, 2015b). Many Member States are still fully dependent on Russian gas imports which amount to 45% of gas imports overall in the second quarter of 2015. Further market integration is likely to improve the robustness in the coming years.

6.5.2.5 *Acceptance and Social Implications*

A recent Eurobarometer poll shows 72% of Europeans are in favour of a common energy policy, the highest support being in Lithuania (84%) and Germany (83%), and considerably lower in the Czech Republic (56%) and Austria (60%) (European Commission, 2015e). Even though this is certainly reassuring, the question may have been formulated too broadly and the topic may be too complex to really yield meaningful results in a standard survey.

It is also worth noting that energy and environment related concerns are not at the top of the agenda: Only 4% of respondents named energy supply as one of the two most important issues facing the EU, 5% the environment and 6% climate change (European Commission, 2015e). These low relevance values are not problematic per se, but signify that the vast majority are focusing on other issues. Values regarding climate and environment are somewhat higher in Denmark, Sweden and Finland.

Asked directly, 92% of Europeans see climate change as a serious problem; 81% believe that fighting climate change and increasing energy efficiency offers economic opportunities for jobs and growth in Europe; and 65% agree that reducing fossil fuel imports will benefit the European economy with the highest approval in Portugal (84%) and Italy (76%) and the lowest in Estonia (35%) and Latvia (45%), indicating clear regional differences and indicating a lack of unity on the issue of energy and climate change in the EU (European Commission, 2015d).

Energy costs are felt particularly by low-income households: 10.2% of Europeans were unable to keep their home adequately warm in 2014, with the highest shares in Bulgaria (40.5%), Greece (32.9%) and Portugal (28.3%) (Eurostat, 2015d).

Energy poverty is the result of low income and high energy consumption prices. 24.4% of Europeans were at risk of poverty or social exclusion in 2014, with values reaching 40.1% in Bulgaria (Eurostat, 2015e).

Populations at risk of poverty are currently not in the focus of EU energy policy. It is noteworthy that Member States with high poverty rates show lower commitment to the European energy transition and Member States with low poverty rates tend to be more ambitious in terms of climate action and environment protection.

6.6 CONCLUSION

COP21 held in Paris in December 2015 is over. Now it is back to the hard work of fighting for, and implementing, the energy transition.

This chapter shows, with particular regard to renewable energy, that in Europe this fight is in progress both at the local level and at the national level. At this level, Germany has arguably achieved better results largely because it has a plan – the *Energiewende*. It is not a perfect plan (as seen) but it is better than nothing, and the effort is off to a good start.

At the European level, the European Commission is far from having the authority to set the national energy policies for each of its Member States. However, the EU set up a framework for bringing Europe's energy policies and infrastructure in line with the overarching security, climate and efficiency agendas (Energy Union). Moreover, Europe is a world leader in renewable energy and has installed three times the amount of renewable power per capita than the rest of the world. Therefore the EU played a core role in the COP21 and will contribute substantially to the implementation of the reached agreement. To put into practice an enlarged conception of 'human energy' thanks to the valorisation of personal and social energy should strengthen this process.

REFERENCES

AG Energiebilanzen, 2015. Bruttostromerzeugung in Deutschland Ab 1990. AG Energiebilanzen. Available from: http://www.ag-energiebilanzen.de/index.php?rex_img_type=rex_220&rex_img_file=20151112_brd_stromerzeugung1990-2014.pdf.

Agentur für Erneuerbare Energien, 2014. RENEWS KOMPAKT Nr. 23/Dezember 2014-Die Ergebnisse der AEE-Akzeptanzumfrage 2014 auf einem Blick. Available from: http://www.unendlich-viel-energie.de/media/file/383.AEE_RenewsKompakt_23_Akzeptanzumfrage2014.pdf.

Agentur für Erneuerbare Energien, 2015a. Energiewende Ist Bürgerwende. Available from: http://www.unendlich-viel-energie.de/themen/akzeptanz2/buergerbeteiligung/energiewende-ist-buergerwende.

Agentur für Erneuerbare Energien, 2015b. Akzeptanz: Blitzumfrage unter Energie-Kommunen. Available from: http://www.unendlich-viel-energie.de/mediathek/grafiken/akzeptanz-blitzumfrage-unter-energie-kommunen.

BDEW, 2015. Wettbewerb 2015. BDEW Bundesverband der Energie- und Wasserwirtschaft e.V. Available from: www.bdew.de/internet.nsf/res/B30B944DA303762EC1257EC7004B032D/$file/150828-BDEW-Wettbewerb-Dt-Energiemarkt-2015-WEB_final.pdf.

Bundesnetzagentur, 2015. Monitoring Report 2014. Bundesnetzagentur. Available from: http://www.bundesnetzagentur.de/SharedDocs/Downloads/EN/BNetzA/PressSection/ReportsPublications/2014/MonitoringReport_2014.pdf?__blob=publicationFile&v=2.

CEER, 2015. CEER Benchmarking Report 5.2-CoS. Available from: http://www.ceer.eu/portal/page/portal/EER_HOME/EER_PUBLICATIONS/CEER_PAPERS/Electricity/Tab4/C14-EQS-62-03_BMR-5-2_Continuity%20of%20Supply_20150127.pdf.

Crivello, S., Karatayev, L., Lombardi, P., Cotella, G., Santangelo, M., Amerighi, O., Slawinski, A, 2013. Deliverable 1.3 Report on main trends in European energy policies. Retrieved from http://www.milesecure2050.eu/documents/public-deliverables/en/deliverable-1-3-report-on-main-trends-in-european-energy-policies.

EEX, 2015. EEX-teilnehmerliste. Available from: www.eex.com/de/handel/teilnehmerliste#/filter.

ENTSOE, 2015. The Successful Stress Test for Europe's Power Grid – More Ahead.

European Commission, 2009. Directive 2009/28/EC of the European Parliament and of the Council of 23 April 2009 on the Promotion of the Use of Energy from Renewable Sources. Available from: http://eur-lex.europa.eu/legal-content/EN/TXT/PDF/?uri=CELEX:32009L0028&from=EN.

Eurostat, 2015f. Electricity prices for domestic consumers - bi-annual data (from 2007 onwards). Available from: http://ec.europa.eu/eurostat/en/web/products-datasets/-/NRG_PC_204.

Eurostat, 2015g. Electricity prices for industrial consumers - bi-annual data (from 2007 onwards). Available from: http://ec.europa.eu/eurostat/en/web/products-datasets/-/NRG_PC_205.

European Commission, 2013. Communication from the Commission to the European Parliament, the Council, the European Economic and Social Committee and the Committee of the Regions: Long Term Infrastructure Vision for Europe and Beyond. Available from: http://eur-lex.europa.eu/resource.html?uri=cellar:a2574790-34e9-11e3-806a-01aa75ed71a1.0007.01/DOC_1&format=PDF.

European Commission, 2014a. A Policy Framework for Climate and Energy in the Period from 2020 to 2030 [COM(2014) 15]. Available from: http://eur-lex.europa.eu/LexUriServ/LexUriServ.do?uri=CELEX:52014DC0015:EN: NOT.

European Commission, 2014b. Energy Prices and Costs in Europe. Available from: https://ec.europa.eu/energy/sites/ener/files/publication/Energy%20Prices%20and%20costs%20in%20Europe%20_en.pdf.

European Commission, 2014c. European Energy Security Strategy COM (2014) 330 Final. European Commission. Available from: http://eur-lex.europa.eu/legal-content/EN/TXT/PDF/?uri=CELEX:52014DC0330&from=EN.

European Commission, 2014d. In-depth Study of European Energy Security [SWD(2014)330]. Available from: https://ec.europa.eu/energy/sites/ener/files/documents/20140528_energy_security_study.pdf.

European Commission, 2015a. Quarterly Report on European Electricity Markets Q2 2015. Available from: https://ec.europa.eu/energy/sites/ener/files/documents/quarterly_report_on_european_electricity_markets_q2_2015.pdf.

European Commission, 2015b. Quarterly Report on European Gas Markets – Q2 2015. Available from: https://ec.europa.eu/energy/sites/ener/files/documents/quarterly_report_on_european_gas_markets_q2_2015.pdf.

European Commission, 2015c. Renewable Energy Progress Report COM(2015) 293 Final.

European Commission, 2015d. Special Eurobarometer 435 Climate Change.

European Commission, 2015e. Standard Eurobarometer 83 Spring 2015.

European Commission, 2015f. State of the Energy Union 2015 COM(2015) 572 Final.

European Commission, 2015g. A Framework Strategy for a Resilient Energy Union with a Forward-Looking Climate Change Policy.

European Commission, 2015h. Climate Action Progress Report.

Eurostat, 28 September 2015a. Electricity Prices by Type of User. Available from: http://ec.europa.eu/eurostat/tgm/table.do?tab=table&init=1&language=en&pcode=ten00117&plugin=1.

Eurostat, 28 September 2015b. Gas Prices by Type of User. Available from: http://ec.europa.eu/eurostat/tgm/graph.do?tab=graph&plugin=1&pcode=ten00118&language=en&toolbox=data.

Eurostat, 28 September 2015c. Market Share of the Largest Generator in the Electricity Market. Available from: http://ec.europa.eu/eurostat/tgm/graph.do?tab=graph&plugin=1&language=en&pcode=ten00119&toolbox=type.

Eurostat, 27 November 2015d. Inability to Keep Home Adequately Warm [ilc_mdes01]. Available from: http://appsso.eurostat.ec.europa.eu/nui/show.do.

Eurostat, 27 November 2015e. People at Risk of Poverty or Social Exclusion by Age and Sex [ilc_peps01]. Available from: http://ec.europa.eu/eurostat/web/income-and-living-conditions/data/database#.

Federal Ministry for Economic Affairs and Energy, August 2014. Das Erneuerbare-Energien-Gesetz 2014-Die wichtigsten Fakten zur Reform des EEG. Available from: http://www.bmwi.de/BMWi/Redaktion/PDF/Publikationen/das-erneuerbare-energien-gesetz-2014,property=pdf,bereich=bmwi2012,sprache=de,rwb=true.pdf.

Federal Ministry for Economic Affairs and Energy, 2015a. Making a Success of the Energy Transition (p. 32). Available from: http://www.bmwi.de/English/Redaktion/Pdf/making-a-success-of-the-energy-transitionproperty=pdf,bereich=bmwi2012,sprache=en,rwb=true.pdf.

Federal Ministry for Economic Affairs and Energy, 2 October 2015b. Quality of Supply Is Improving in Germany. Federal Ministry for Economic Affairs and Energy. Available from: www.bmwi-energiewende.de/EWD/Redaktion/EN/Newsletter/2015/10/Meldung/infografik-quality-of-supply-improving.html.

Federal Ministry for Economic Affairs and Energy, 2015c. Fourth Energy Transition Progress Report. Available from: http://www.bmwi.de/English/Redaktion/Pdf/vierter-monitoring-bericht-energie-der-zukunft-kurzfassung,property=pdf,bereich=bmwi2012,sprache=en,rwb=true.pdf.

German Federal Government, 2015. Bilanz zur Energiewende 2015. Available from: http://www.bundesregierung.de/Content/DE/_Anlagen/2015/03/2015-03-23-bilanz-energiewende-2015.pdf?__blob=publicationFile&v=1.

German Federal Government, 2010. Energiekonzept für eine umweltschonende, zuverlässige und bezahlbare Energieversorgung. Available from: http://www.bundesregierung.de/ContentArchiv/DE/Archiv17/_Anlagen/2012/02/energiekonzept-final.pdf;jsessionid=915CE03A123028E6A697E477F7E54F58.s2t1?__blob=publicationFile&v=5.

German Federal Ministry for Economic Affairs and Energy, 2014. Die Energie der Zukunft: Erster Fortschrittbericht der Energiewende. German Federal Ministry for Economic Affairs and Energy. Available from: http://www.bmwi.de/BMWi/Redaktion/PDF/Publikationen/fortschritsbericht,property=pdf,bereich=bmwi2012,sprache=de,rwb=true.pdf.

GME, 2014. Relazione Annuale 2014. Gestore dei Mercati Elettrici.

GSE, March 2015. Rapporto statistico Energia da fonti rinnovabili 2013. Gestore dei Servizi Elettrici.

Italian Ministry of Economic Development, March 2013. Italy's National Energy Strategy: For a More Competitive and Sustainable Energy. Available at: http://www.encharter.org/fileadmin/user_upload/Energy_policies_and_legislation/Italy_2013_National_Energy_Strategy_ENG.pdf.

IW Consult, 2014. BDI-Energiewende-Navigator 2014 – Monitoring zur Umsetzung der Energiewende. Available from: http://www.iwkoeln.de/_storage/asset/201041/storage/master/file/5662779/download/bdi-energiewende-navigator_2014.pdf.

Legambiente, 2015. Rapporto Comuni Rinnovabili 2015.

MILESECURE-2050, December 2013. Deliverable 1.3, "Report on Main Trends in European Energy Policies". Available at: http://www.milesecure2050.eu/en/public-deliverables.

Regelleistung.net, 10 April 2014. Information on Grid Control Cooperation and International Development. Available from: www.regelleistung.net/ext/download/marktinformationenApgEn.

Stark, M., September 2015. System and Market Integration – the Perspective of a German Energy Trader. Berlin.

World Economic Forum, 2015. The Global Competitiveness Report 2015–2016. World Economic Forum. Available from: http://www3.weforum.org/docs/gcr/2015-2016/Global_Competitiveness_Report_2015-2016.pdf.

Chapter 7

Energy Security in Low-Carbon Pathways

C. Cassen[1], F. Gracceva[2]

[1]CIRED (Centre International de Recherche sur l'environnement et le développement), CNRS (Centre national de la recherche scientifique), Nogent sur Marne Cedex, France; [2]Studies and Strategies Unit, ENEA (Italian National Agency for New Technologies, Energy and Sustainable Economic Development), Rome, Italy

7.1 INTRODUCTION

The intellectual debates expressed by the Club of Rome about the 'Limits to Growth' (Meadows et al., 1972) and the oil crisis in the 1970s highlighted for the first time the close interactions between environmental, economic development and energy security issues. The international community has since turned more of its attention to these interactions. For example, energy security issues were one of the main drivers that put a climate convention on the international agenda after the G7 meeting at Houston in 1990, on George H. W. Bush's initiative (Kirton, 2007; Guivarch et al., 2015). The rise of climate change on the public agenda since the late 1980s has also prompted the need for the quantitative assessment of the costs and impacts of mitigation strategies. To meet this demand, an increasing number of scenarios have been produced by energy-economy-environment (E3) models which represent collection of different types of models, in particular the so-called integrated assessment models (IAMs). This demand was in part mediated by the IPCC, which developed at the nexus of climate research and politics.[1]

This chapter provides an overview of the main findings provided by the global scenarios synthesised in the Working Group III of the fifth assessment report (AR5) of the IPCC (2014). These scenarios produced by IAMs emphasise the very limited room for manoeuvre in order to meet the target of a global mean surface temperature increase below 2°C (which has been an official target of climate negotiations since the Copenhagen conference in 2009), notably if

---

1. The IPCC was established after the Toronto Conference in 1988. It is composed of three working groups. Working Group I deals with climate science, Working Group II deals with the impacts of climate change and strategies of adaptation and Working Group III addresses climate mitigation policies.

Low-carbon Energy Security from a European Perspective. http://dx.doi.org/10.1016/B978-0-12-802970-1.00007-3
Copyright © 2016 Elsevier Ltd. All rights reserved.

ambitious policies to reduce greenhouse gases (GHGs) are not implemented by 2020. This also demands deep transformations in energy systems, both at multilevel scales (regional, national and sectoral level), and in terms of people's lifestyles which, in turn, raise challenges for energy security.

The first section provides a review of the global scenarios in the fifth IPCC report and analyses the conditions necessary for a 2°C target. The second section focuses on the global evaluation of energy security challenges in low-carbon pathways. This section presents a critical overview of the literature on the interactions between energy security and climate change policies and addresses the increasing interest of IAMs to assess the co-benefits of climate policies primarily based on the body of work of the Global Energy Assessment (GEA).[2] It also shows how the current process of climate negotiations in the wake of the Paris Agreement can open a window of opportunity for multiobjective policies which seek synergies between climate policies and other development objectives (poverty alleviation, employment, energy security, health, etc.) and considers the methodological implications imposed on IAMs to assess co-benefits.

7.2 REVIEW OF GLOBAL SCENARIOS IN THE FIFTH IPCC REPORT

7.2.1 The Scenarios Database Provided by Integrated Assessment Models

The E3 modelling agenda has been dominated since the fourth assessment report (AR4) published in 2007 by the evaluation of techno-economic implications of stringent carbon trajectories, in particular the 2°C target. Chapter 6 of the AR5 Working Group III is dedicated to assessing 'transformation pathways' (Clarke et al., 2014) and reviewed data from 1184 new socioeconomic scenarios published since 2007.[3] These scenarios were generated primarily by large-scale IAMs that can project key characteristics of transformation pathways to 2050

---

2. Coordinated by IIASA and launched in 2012 for Rio+20 and gathering 300 experts covering a large set of disciplines, GEA pathways explore some 40 pathways that satisfy simultaneously the normative, social and environmental goals: stabilising global climate change to 2°C above preindustrial levels to be achieved in the 21st century, enhancing energy security, eliminating air pollution and reaching universal access to modern energy services by 2030. Five hundred independent experts from academia, business, government, intergovernmental and nongovernmental organisations from ·all regions of the world have contributed to GEA in a process similar to the IPCC. Long-term energy and economy pathways are based on the IAMs IMAGE and MESSAGE models (see http://www.iiasa.ac.at/web/home/research/Flagship-Projects/Global-Energy-Assessment/Home-GEA.en.html); the database of scenarios is available at http://www.iiasa.ac.at/web-apps/ene/geadb/dsd?Action=htmlpage&page=about).

3. This reveals an inflation of scenarios, as the third IPCC report included 380 scenarios with 26 models and the fourth IPCC report integrated 780 scenarios. More generally, rising demand for scenarios for research, expertise and support to decision making, and the exponential development of computing power favour the development of models, leading to 100 simulations in a few years.

and beyond. IAMs are simplified, stylised numerical approaches to represent complex physical and social systems, and the most relevant interactions within the systems (eg, energy, agriculture, the economic system). They take in a set of input assumptions and produce outputs such as energy system transitions, land-use transitions, economic effects of mitigation, and emissions trajectories (Sarofim and Reilly, 2011). Important input assumptions include population growth, baseline economic growth or total factor productivity growth, fossil fuel resources, technology costs or learning rates (see also Box 7.1 for a detailed presentation of the community of IAMs).

The landscape of IAMs is also divided between predominant energy-technology models and macroeconomic models, the former being called bottom-up (BU) models and the latter top-down (TD) models, as a result of attempts to classify models inspired by Zhang and Folmer (1998), Chapter 7

Box 7.1 The Community of Integrated Assessment Models

There are around 30 integrated assessment models (IAMs) in the world. These models have been developed within research institutes (originally, mainly in the Organisation for Economic Cooperation and Development (OECD) countries before spreading to countries such as India, China, South Africa and Brazil), international institutions (OECD and International Energy Agency (IEA)) or governmental structures (such as the PBL Netherlands Environmental Assessment Agency or the Australian Bureau of Agricultural and Resources Economics). Within the OECD countries, leading research institutes are located in the United States (Pacific Northwest National Laboratory (PNNL), Massachusetts Institute of Technology (MIT) and Stanford University), in Europe (Potsdam Institute for Climate (PIK), Fondazione Eni Enrico Mattei (FEEM), PBL Netherlands Environmental Assessment Agency, International Institute for Applied Systems Analysis (IIASA), Centre international de recherche sur l'environnement et le développement (CIRED) and Edden/University of Grenoble), in Australia and in Japan (National Institute for Environmental Studies). The community of IAMs is organised around research networks such as the Energy Modelling Forum (EMF), the Integrated Assessment Modelling Consortium (IAMC) at Stanford University, the Energy Technology System Analysis Programme (ETSAP), the IIASA in Vienna and intercomparison modelling exercises funded by the sixth and seventh European Union (EU) Framework Programmes (EU FP6 and FP7). Jointly with the EMF modelling exercises,[4] EU FP7 projects[5] provided a large part of socioeconomic scenarios analysed in the fifth IPCC report.

---

[4] Coordinated by Stanford University, EMF organised several modelling workshops since the AR4. EMF22 (Clarke et al., 2009) assessed (1) the feasibility of low stabilisation scenarios given an emission reduction target, whether or not this target can be temporarily exceeded prior to 2100 ('overshoot') allowing for greater near-term flexibility, and (2) the nature of international participation in emissions mitigation. EMF27 dealt with the feasibility of low-carbon trajectories consistent with the 2°C target and the low-carbon options to comply with this objective (Weyant et al., 2014).
[5] This refers to EU FP6 project ADAM (http://www.adamproject.eu/), EU FP7 projects AMPERE (http://ampere-project.eu/web/), LIMITS (http://www.feem-project.net/limits/), and to the RECIPE project (https://www.pik-potsdam.de/recipe-groupspace/).

of the third IPCC report Markandya et al. (2001), Crassous (2008), Guivarch (2009) and Amerighi et al. (2010).

BU, or energy systems models, contain a lot of technological and sectoral information expressed in a language similar to the one used by engineers (eg, requirement for heat insulation, lighting, mobility expressed in physical quanti-ties). The economic component is reduced to the projection of basic economic growth scenarios (gross domestic product [GDP], GDP sectorial allocation and household incomes) and to assumptions on technology and fossil fuel costs. A prominent BU model is the MARKAL model.[6]

Conversely, TD models analyse principally economic interdependencies among social groups, sectors and countries. They use the economics tool box derived from Von Neuman's general equilibrium model to represent flows of values but rely on a rough representation of the technological content of scenar-ios as they encapsulate a limited description of technologies. Inside this 'com-munity', three types of TD models or macroeconomic models go together:

- Optimal growth models, which are built on the principle of intertemporal maximisation of one single representative agent.
- Macroeconometric models, which project future scenarios from econometric relations between economic variables calibrated on past data.
- Computable general equilibrium models (CGE), which represent all markets and their interdependencies as well as all the budget equations of representative agents.

The gap between TD and BU approaches has narrowed rapidly with the increasing number of hybrid models characterised by a comprehensive TD rep-resentation of the macroeconomic processes and complemented by a techno-logically explicit BU representation of energy systems. BU models have been integrated with CGE modules by internal or external iterative convergence pro-cess (eg, ETA MACRO, MARKAL MACRO). This has been done through soft or hard linking modes (Crassous, 2008; Bibas, 2015). Although the progress and challenges of hybrid modelling have been addressed in a special issue of *The Energy Journal* (Hourcade et al., 2006), hybrid modelling has to continue to develop and improve (Bibas, 2015).

The next section presents the main findings of the scenarios provided by IAMs in the fifth Assessment Report in greater depth.

7.2.2 Which Low-Carbon Trajectory is Compatible With the 2°C Target?

The 5th IPCC report shares a pessimistic diagnosis of GHG emissions trends over the last decade. Whilst efforts to reduce GHG emissions were implemented

---

6. The MARKAL model is based on linear programming to optimise primary energy supply needed to meet a given level of final demand.

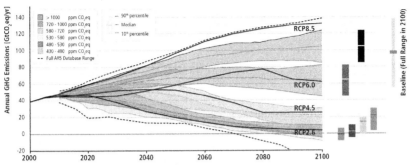

FIGURE 7.1 Greenhouse gas (GHG) emissions trajectories from 2000 to 2100 (GtCO₂eq/an) given by scenarios. *Reproduced from: Summary for Policy makers (SPM), IPCC, 2014. Climate Change 2014: Mitigation of Climate Change: Contribution of Working Group III to the Fifth Assessment Report of the Intergovernmental Panel on Climate Change. Edenhofer, O., Pichs-Madruga, R., Sokona, Y., Farahani, E., Kadner, S., Seyboth, K., Adler, A., Baum, I., Brunner, S., Eickemeier, P., Kriemann, B., Savolainen, J., Schlömer, S., von Stechow, C., Zwickel, T., Minx, J.C. (Eds.). Cambridge University Press, Cambridge (UK) and New York.*

following the entry into force of the Kyoto Protocol in 2005, GHG emissions accelerated from 1.4%/year between 1970 and 2000 to 2.2%/year between 2000 and 2010. This acceleration was primarily triggered by the rising economic growth in emerging countries which have become the main drivers of emissions, and have overcome the efforts in GDP energy intensity gains (measured by the reduction of energy used per unit of GDP) at the global level. This breaks with the continuous decrease of energy carbon intensity (quantity of CO_2 per unit of energy used) that has been observed over the last decades.

Projections of GHG emissions by IAMs cover a range comparable to those of the four representative concentration pathways (RCP)[7] analysed by the IPCC Working Group I (Fig. 7.1). Therefore, it is worthwhile to study the database of scenarios to address some key questions: (1) What would be the magnitude of climate change derived from scenarios without supplementary climate mitigation efforts? (2) What emissions trajectories can reach the 2°C official target? and (3) What are the subsequent necessary transformations required in both the energy systems and on the demand side to comply with this objective?

IPCC scenarios emphasise the necessity to implement additional mitigation efforts to avoid the risk of an increase in the average global temperature of between 3.7°C and 4.8°C in 2100 based on the preindustrial period. In scenarios with a probability to remain under the threshold of 2°C superior to 50%, the GHG concentration in 2100 is close to 450 ppm eq CO_2. These scenarios also

---

7. RCP include four GHG concentration trajectories, each RCP corresponds to the gradation forcing reached in 2100: RCP2.6 corresponds to radiation forcing of d +2.6 W/m², scenario RCP4.5 to +4.5 W/m², scenario RCP6 to +6 W/m² and scenario RCP8.5 to +8.5 W/m². These scenarios were used as a basis for future climate change projections works.

insist on the increased risk of failing to meet the 2°C target if the implementation of mitigation actions is delayed unless it achieves increased negative emissions as it is shown in the following section.

7.2.3 Deep Transformations Required in Energy Systems

Low-carbon trajectories in IPCC scenarios are triggered by major technical and institutional changes. On the supply side, this is characterised by high energy efficiency gains (see Section 7.2.4) and a high deployment of zero- or low-carbon energies (renewables, nuclear, fossil fuels with carbon capture and storage) or negative emissions (bioenergy with carbon sequestration).

The share of renewables in the energy mix in scenarios compatible with the 2°C target is multiplied by four by 2050 from 15% to 60% (Fig. 7.2) and by 2100, supplies of low-carbon energy – energy from nuclear power, solar power, wind power, hydroelectric power, bioenergy and fossil resources with carbon dioxide capture and storage – might need to increase five-fold or more over the next 40 years (Clarke et al., 2014).

It is worth noting that this represents a deep and ambitious transformation in the energy mix. In 2013, about 80% of global primary energy supply still came from fossil fuels which also represent 78% of gross EU energy consumption (Eurostat, 2015). Biomass, which is mostly used for heating, was globally the main renewable energy source. The current level of bioenergy use has signalled opportunities for negative emissions to help meet the 2°C target. However, this approach is not without its problems.

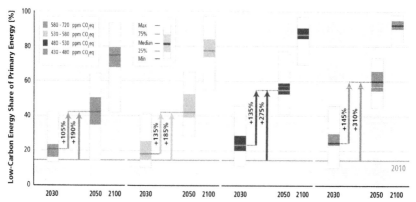

FIGURE 7.2 Changes in share of low-carbon energies (expressed in % of primary energy) for 2030, 2050 and 2100 with respect to 2010 in mitigation scenarios identified by the array of CO_2-eq concentration reached in 2100. *Reproduced from: Summary for Policy makers (SPM), IPCC, 2014. Climate Change 2014: Mitigation of Climate Change: Contribution of Working Group III to the Fifth Assessment Report of the Intergovernmental Panel on Climate Change. Edenhofer, O., Pichs-Madruga, R., Sokona, Y., Farahani, E., Kadner, S., Seyboth, K., Adler, A., Baum, I., Brunner, S., Eickemeier, P., Kriemann, B., Savolainen, J., Schlömer, S., von Stechow, C., Zwickel, T., Minx, J.C. (Eds.). Cambridge University Press, Cambridge (UK) and New York.*

Achieving negative emissions indeed forced IAMs to assume ambitious hypotheses in terms of deployment of low-carbon technologies at a large scale, associated with negative emissions technologies and accelerating afforestation. Bioenergy with carbon capture and storage (BECCS) technology has been introduced in most of IAMs to comply with the 2°C target. In scenarios developed for instance within the EU FP7 Adaptation and Mitigation Strategies Initiative (ADAM) project, the use of bioenergy in combination with BECCS would amount up to 400 EJ/yr used at its maximum for the lowest stabilisation target (400 ppm CO_2 eq) (Van Vuuren et al., 2010).

However, these assumptions pose a set of sensitive issues that relate to the competition for land use and water necessary for biomass development, the impacts on agricultural prices and the uncertainties around the risks associated with carbon storage (Benson, 2014; van der Meer et al., 2014). Based on Haberl et al. (2010) which assumed yields with energy equivalent of 149 GJ/ha/year (in a context of 'managed forests'), Bibas and Mejean (2014) show that the production of 300 EJ/year of biomass worldwide would require a land area of 2.01 Gha, corresponding to about 15% of total land area.[8] Uncertainties remain at least on the technical feasibility and social acceptability of a large scale development of BECCS, in particular regarding the size of geological storage capacity. Therefore, achieving negative emissions at a global scale can be a high-risk strategy (Fuss et al., 2014).

Whilst IPCC scenarios offer a thorough description of a future energy mix through a technology centric approach, they deliver few elements regarding the necessary changes on the demand side. Changes in consumption behaviours and patterns (eg, in terms of transport and diet, are deemed necessary to support the structural changes of the economy. However, changes in diets and reduction of food wastage have yet to be explored in depth, although there are substantial emission reduction potentials. Tilman and Clark (2014) show that the development of diets high in fats, meat and sugar by 2050 would mainly contribute to estimated 80% increase in global agricultural GHG emissions from food production and to global land clearing. Alternative diets could reduce global agricultural GHG emissions and provide substantial health benefits.

7.2.4 Sectoral Potentials of the Low-Carbon Transition

The set of IPCC scenarios also shows that achieving a 2°C target requires deep transformation at the sectoral level. Power supply is the sector that experiences the most rapid process of decarbonisation compared to industry, transport and building. The reduction of the carbon content of each unit of electricity produced is followed by an electrification of end use final services.

---

8. See also Brunelle (2012) and Souty (2013) for a review of the controversies on land use, land-use change and forestry in models and Lambin and Meyfroidt (2011) for an analysis of the drivers of land availability.

In mitigation scenarios, the share of low-carbon technologies in electricity production (renewables, nuclear and carbon capture and storage [CCS]) is between 30% and 80% in 2050 and accounts for nearly 100% in 2100. These trends follow the spectacular progress of performance and cost decrease accomplished over the last years by several renewable energy production technologies (for instance, wind and photovoltaic). In 2012, renewable energy represented more than 50% of newly installed electricity capacities on the global scale. Large scale deployment is, however, still dependent on direct or indirect public supports should a redeployment of fossil fuel subsidies takes place.

Whilst transport represented 27% of final world energy needs in 2010 and accounted for $6.7 \, GtCO_2$ in 2010, these emissions could double by 2050 without mitigation efforts, mainly because of the rising of mobility needs (air and terrestrial) on a global scale. Sectoral studies suggest that significant potential for emissions reductions exist in the transport sector. In the short term, higher energy performance gains by cars (20–30% by 2050) are counterbalanced by increasing mobility, in particular in developing countries. But in the long term, the rising penetration of zero-carbon vehicles jointly associated with a reorganisation of transport infrastructures and urban forms could limit the rising demand of transport and increase the potentials of emissions reductions in scenarios that comply with the 2°C target.

Emissions from the building sector have doubled over the last 40 years ($8.8 \, GtCO_2$). Energy demand could be multiplied by two by 2050, triggered by greater global urbanisation and lifestyles in developing countries catching up to those of developed countries (eg, rising size of flats and household equipment). The deployment of low-energy consumption options (insulation, lighting, etc.) could at least help to stabilise energy demand by 2050.[9]

GHG emissions from the industry sector correspond to 30% of global emissions ($13 \, GtCO_2$ in 2010). Gains in energy intensity are estimated at 25% thanks to the deployment of 'best practices' for industrial processes (eg, increase in energy efficiency, changes in choices of raw materials, recycling of materials, setting carbon sequestration systems, radical shifts in products design and production, in particular in the production of cement[10]).

Although the emissions of the agricultural and forest sector have stabilised during the last decade, there is still additional supplementary potential based on

---

9. Recent advances in technologies, appliances design and know-how associated with sober modes of consumption can reduce energy needs of new buildings by a factor of 2–10, with moderate and negative costs (eg, heating and lighting and in commercial buildings). However, a series of barriers hinder the implementation of low-carbon measures: long return on investment regarding the refurbishing and construction of buildings, asymmetrical relationships between owners and renters or difficult access to specific information related to energy economies.

10. The production of cement is the source of sizeable carbon dioxide emissions. Final energy reduction for industry relative to baseline scenarios is between 22% and 38% interquartile ranges for the whole sector in 2050 with 430–530 ppm CO_2-eq concentrations in 2100. Energy efficiency improvements are the results of new processes/technologies (Clarke et al., 2014).

the fight against deforestation and increased forest management, the introduction of cultural practices (generalisation of direct sewage for instance) and less emitting pasture. The development of bioenergy is strong although its impacts on land use (land availability, potential competition with food needs, pollution) need to be further analysed, as mentioned in the previous section.

From this brief overview of sectoral challenges related to mitigation, it appears that significant emission reduction potentials exist. However, the lifespan of the infrastructures and the related inertias have to be taken into account. Indeed, in all sectors, equipment and long-lived infrastructures (one or two decades for cars, several decades for industrial installed capacities or power plant units, several decades for buildings, roads, etc.) impact energy consumptions and increase the cost of transformations. Investment choices today in these infrastructures will impact the feasibility and the cost of mitigation actions in the long term. These scenarios also show potential challenges for energy security on the global and regional level. For instance, the higher penetration of RES raises concerns about their integration in the power supply. Indeed, their intrinsic unpredictability and variability to produce a certain amount of energy over time scales ranging from seconds to years potentially represents an internal stress for the energy system which characterises one dimension of energy security. This can 'constrain the ease of integration and result in additional system costs, particularly when reaching higher RE shares of electricity, heat or gaseous and liquid fuels' (IPCC, 2011). Higher electricity prices in the short term resulting from carbon pricing policies could also negatively affect the poorest households despite energy savings.

The question that follows is to what extent climate policies can provide coherent answers to these issues. Section 7.3 will show that the assessment of the synergies and trade-offs with other key dimensions of the transition, among them energy security issues, is a key issue for models. This increasing interest in IPCC for co-benefits is also closely related to the current evolution of the climate negotiations process.

7.3 GLOBAL EVALUATION OF ENERGY SECURITY CHALLENGES IN LOW-CARBON PATHWAYS

Security of supply, sustainability and competitiveness are the three complementary pillars/goals of the European energy policy (EC, 2006) and as such they are indeed 'part of the same strategy. Work to achieve one should help deliver the others' (EC, 2010). However, the EU Commission itself has recently recognised the risk that climate-focused energy policies, if not properly designed, can affect energy security and bring about extra costs, as they support technological and market solutions designed to achieve a different policy objective (EC, 2013).

In fact, the wide range of policies directed to design an EU low-carbon energy system have often been introduced assuming that security of supply, sustainability and competitiveness are part of the same strategy (EU, 2006, 2010).

There are indeed obvious synergies between the different targets, but there are also potential trade-offs. At the same time the radical changes envisaged by these policies are now causing new challenges for the future energy security of supply both in the short and medium term, notably in some EU electricity markets. On one hand, the increased penetration of Variable Energy Resources adds a set of specific operating challenges, which makes it difficult to guarantee the stability of the power system (Baritaud, 2012). On the other hand, the current EU power market seems unable to value the benefits of flexible resources. Indeed, thermal flexible plant operators are finding it increasingly difficult to recover their fixed costs, due to the combination of persistently low market prices brought about by renewable plants (with virtually zero marginal costs) setting market clearing prices, and of the much lower load factors of thermal plants (Haas et al., 2013). Moreover, relatively costly European gas and persistently inexpensive coal (exacerbated by a weak carbon price) has placed competitive strains on natural gas-fired power plants, along with the large utilities operating them. The resulting low profitability of natural gas-fired plants reduces the incentive to invest in infrastructure upgrades needed to realise a system in line with political goals. In order to tackle these challenges, the EC 2030 framework for climate and energy stresses that 'the 2030 framework must identify how best to maximize synergies and deal with trade-offs between the objectives of competitiveness, security of energy supply and sustainability' (EC, 2013).

A key challenge in meeting this objective lies in the different 'natures' of energy security and climate change mitigation. Because GHG mitigation is measurable in a relatively straightforward way, it is clear whether or not policies are heading in the right direction. On the other hand, even without being set in the context of sustainability, security of energy supply is an inherently complex topic: as often underlined in the literature, much of the discussion is 'conducted without a clear idea of the dimensions of energy security and their relative significance' (Watson, 2009; Chester, 2010; Loschel et al., 2010; Hughes, 2009). As consequence of the complexity of the topic, most existing assessments of the links between climate change and energy security present several methodological deficiencies. The next section provides a critical overview of these assessments and their limitations.

7.3.1 Interactions Between Energy Security and Climate Change Policies: A Review

The debate on the interaction between energy security and climate change policies is typically framed in terms of the trade-offs and synergies between the two policies. Many studies argue that optimal policies, which can mitigate climate change and enhance security at the same time, are possible (eg, Brown and Huntington, 2008; Greenleaf et al., 2009; Criqui and Mima, 2011). Others contend that the two goals are often fundamentally at odds (Luft and Korin, 2011). Evidence can be presented for both sides (Watson, 2009). At the

European political level, considerations of energy security have sometimes trumped climate change commitments (demonstrated by Poland's opposition to switch from domestic coal to imported gas) and sometimes not (witnessed by the acceptance of Germany's *Energiewende* despite the acknowledgement of higher energy costs and greater risks of grid instability).

Framing the energy security/climate change policy nexus in terms of synergies and trade-offs creates a compelling case for quantification. But whereas the essence of climate change mitigation policies can indeed be reduced to a desire to control the amount of carbon emitted by humans into the atmosphere, energy security – given its conceptually elusive and multidimensional nature – is not so easily reducible to a single property. The consequence is that the existing literature presents several shortcomings.

Indeed, a significant number of studies purporting to explore the interactions between climate and energy security policies end up evaluating the impact of climate policies on only a subset of factors that may or may not capture the essence of energy security (eg, Wu et al., 2012). This is because most studies are typically based on the combination of a model-based scenario analysis with a set of indicators. Designed to reduce complex phenomena to simple terms and functions, indicators are widely used to abstract from the energy system a few key input parameters to give an overall indication of its level of security (for an overview see Greenleaf et al., 2009; Kruyt et al., 2009; Martchamadol and Kumar, 2013). Yet despite their ability to simplify complex phenomena, indicators suffer from some key weaknesses that challenge their actual usefulness as policy instruments.

First, energy security indicators are a signal of the state of an energy system, conveying information about its potential vulnerabilities. As such, their essence is to simplify what would otherwise be a complex phenomenon defying quantification, because the complexity of energy systems hides multiple dynamic vulnerabilities (Cherp and Jewell, 2011) and because the security of an energy system depends on a wide number of interrelated factors determining its capacity 'to tolerate disturbance and to continue to deliver affordable energy services to consumers' (Chaudry et al., 2009). By their very nature, indicators are unable to assess the energy system's response to adverse events, ie, the vulnerability of the system in terms of the actual consequences of energy insecurity, despite claims to the contrary (Greenleaf et al., 2009; Jewell, 2011; Jansen et al., 2004). This is because indicator values are not dynamically linked, and hence processes such as primary energy substitutions or demand elasticities go unaccounted for. Indeed, even very sophisticated analyses of the interactions between energy security and mitigation policies are ultimately based on variables which are proxies of the vulnerability of the system (Greenleaf et al., 2009; Augutis et al., 2012).

A direct consequence of these weaknesses is that many indicator-based assessments often focus on diversity as a desired state for energy systems, making the degree of diversification the de facto measure of energy security

(Jun et al., 2009). However, there exist many other dimensions of supply security that extend beyond the issue of diversity alone (Stirling, 2009). Moreover, options to increase diversity means investing in alternatives whose lack of penetration in the energy system may be due to poor performance; thus, enlarging their contributions would imply some penalty (Stirling, 2011). A limit of indicators is indeed that they cannot provide insights on the key issue of the costs and benefits of alternative levels of energy security, which can then be benchmarked against climate targets (DBERR, 2007; Shuttleworth, 2002). Two interesting exceptions in this respect are Maisonnave et al. (2012) and Rozenberg et al. (2010), who use model-based scenario analyses to assess the economic cost of two specific threats, oil price hikes and oil scarcity, with and without a climate policy, as well as the economic costs of not exploiting fossil fuels versus suffering through climate change. Indeed, energy system models have the potential to provide estimations of the economic costs of long-term energy insecurity under different mitigation strategies. Two studies showing the potential of this approach are Babonneau et al. (2009) and the UK study Energy 2050 (Chaudry et al., 2009; Skea et al., 2011). Labriet et al. (2009) focused on the trade-off between the energy system cost for EU and the overall reliability of the EU energy supply, defined as the guaranteed probability of satisfying the energy service demand of the system through random energy import channels. In Chaudry et al. (2009) and Skea et al. (2011) the authors assess a set of strategies aiming at improving the security of the system by comparing the costs of their implementation with their effect on the estimated cost of energy insecurity: they calculate the expected frequency of a crisis at which any strategy breaks even, under different long-term energy system scenarios (including a low-carbon scenario and a resilient scenario).

Finally, the indicator-based approach is particularly prone to view energy security as a distinctly supply-side phenomenon, with its root cause ultimately traced to the risk of a disruption to the smooth functioning of the primary energy supply chain. This is the case of the several studies where the energy security implication of mitigation scenarios are assessed by looking at aspects such as the market concentration in competitive fossil fuel markets and pipeline-based gas import for regulated markets (Lefèvre, 2007); the structure of global oil and gas production and trade (Criqui and Mima, 2011); or a wide set of indicators related to oil and gas resources and production, market concentration and energy trade (Greenleaf et al., 2009; Kruyt et al., 2009). Studies that narrowly focus on supply side-aspects of energy security tend to see positive outcomes of climate change policies since they usually result in the receding dominance of fossil fuels in the supply mix (Lefèvre, 2007; Guivarch et al., 2015). For example, one study attempts to empirically link energy security policies (understood in terms of fossil fuel resource concentration) with carbon emissions in key OECD countries (Lefèvre, 2010). An exception to this narrow approach to energy security is included in Gracceva and Zeniewski (2014, 2015): first, they address the multidimensionality of energy security by identifying a set of five properties of a 'secure' energy system, namely *stability, flexibility, resilience, adequacy* and *robustness*; second, a multiregional energy system model is then used to assess the impact of a low-carbon scenario on these five properties. The results demonstrate how this scenario induces structural changes along the whole energy supply chain, revealing a wide range of implications for the security of the energy system as well as dynamic vulnerabilities and trade-offs.

The methodological flaws discussed so far are also present in the assessments of the interactions between climate policies and energy security developed within the IPCC Working Group III. However, given the prominence of IPCC assessment reports in the climate change mitigation debate, it is worth a more in-depth analysis of the way they assess the co-benefits of climate policies.

7.3.2 A Rising Interest of Co-benefits Analysis in IPCC Assessment Report 5

It is worth noting that whilst most of the modelling exercises related to climate change mitigation policies have principally focused on the feasibility of the 2°C target, using a relative technology-centric approach, the AR5 Group III report represents a step forward, from a qualitative and quantitative perspective, regarding the analysis of co-benefits of climate policies (Cassen et al., 2015). Since the fourth IPCC report (2007), the examination of multiobjectives policies has indeed seen progress in the literature. The number of references of the term 'co-benefit' increased, from 31 citations in the fourth IPCC report to 61 in the technical summary of the fifth IPCC. Sectoral assessments of co-benefits are more systematic and synthesised in a table in Chapters 8 to 12 which are dedicated respectively to transport, building, industry, agriculture and cities.

Co-benefits of climate policies cover a broad and multidimensional perimeter regarding health, environmental impacts, energy access, employment, etc. For instance, replacing a coal plant with a gas plant or by renewables, implementing measures to limit traffic jams in cities or improving insulation in houses can lead to positive impacts on health (reduced GHG emissions and solid particles), energy security (reduced household energy bills) and can additionally foster environmental conservation. Conversely, a negative impact (side effect) could be the increase of energy prices for households and firms resulting from the implementation of a carbon price. The recent review by Ürge-Vorsatz et al. (2014) points out the multifaceted nature of co-benefits by identifying the large set of meanings of this term in the literature and the potential confusions.

This section concentrates more specifically on the synergies and trade-offs between climate policies and energy security. Energy security is multidimensional in nature and IAMs allow for the pursuit of a 'holistic, system wide integrated assessment' of energy security, in particular by assessing the impacts of 'climate mitigation side-effects'. One prominent exercise, in respect to these synergies and trade-offs, was the GEA, whose scenarios are part of the scenario database of IPCC Group III.

7.3.2.1 The Global Energy Assessment

The GEA scenarios shed light on the macrolevel implications of climate mitigation for other societal priorities, including energy security, in addition to energy access, air pollution and its health impacts, water use, land use requirements and biodiversity preservation.

It develops and then assesses, using two IAMs, the IMAGE and MESSAGE models, three long term pathways which describe transformations towards normative objectives for energy access, environmental impacts of energy conversion and use, and energy security (Fig. 7.3). GEA-supply pathways focus on supply-side options based on liquid fuels, gas and optimistic assumptions of the development of CCS whilst in the GEA-efficiency pathways, energy efficiency, a large-scale deployment of RES and high-energy savings are the main drivers of the decarbonisation process. GEA-mix pathways include both dimensions (efficiency and supply). Variants, the so-called branching points, are considered for each group of pathways differentiated by the level of demand (low,

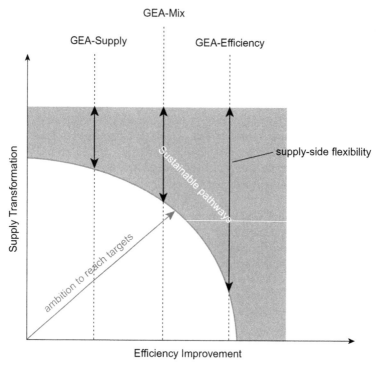

FIGURE 7.3 The Global Energy Assessment pathways. *Reproduced from: Riahi, K., F. Dentener, D. Gielen, A. Grubler, J. Jewell, Z. Klimont, V. Krey, D. McCollum, S. Pachauri, S. Rao, B. van Ruijven, D. P. van Vuuren and C. Wilson, 2012: Chapter 17 - Energy Pathways for Sustainable Development. In Global Energy Assessment - Toward a Sustainable Future, Cambridge University Press, Cambridge, UK and New York, NY, USA and the International Institute for Applied Systems Analysis, Laxenburg, Austria, pp. 1203–1306.*

intermediate, or high), transportation system transformation (conventional or advanced) and a portfolio of supply-side options (full or restricted). In one alternative, the transport system remains conventional, relying predominantly on liquid fuels (including some oil), biofuels, liquefied natural gas, and potentially the direct use of biogas and natural gas without discontinuity from current trends in terms of both end uses technologies, fuel supply and distribution infrastructure. In contrast, an advanced transportation system involves deeper transformations that largely require new infrastructure systems in the case of hydrogen fuel cell vehicles, or new uses for existing infrastructure in the case of plug-in hybrids or fully electric vehicles.

The GEA analysis then considers three dimensions of energy security derived from (Cherp et al., 2012) energy security being defined as 'low vulnerability of vital energy systems'[11]: sovereignty (relating to geopolitics, power balance in energy trade and control over energy systems), resilience (ability to respond to disruption related to price volatility, for instance) and robustness (physical state of infrastructures, impact of possible black out, scarcity of resources and accidents). A set of indicators is associated with each dimension to assess the effect of policies implemented in the three pathways.

Global trade in energy is used as a proxy to measure the sovereignty aspects of energy security. By decreasing energy intensity, ambitious climate policies have the co-benefit to lower oil demand and trade which peaks around 2030 in all the GEA scenarios. However, gas is used as a transition fuel and experiences growth to some 20% of global primary energy supply (compared with oil's 36% share today) in 2050, with increasing trade flows and a decrease in the diversity of production. The decline in absolute trade volumes of fossil fuel after 2030 is most pronounced in the Advanced Transport GEA-Mix and Efficiency pathways. In this context, the GEA analysis concludes that overall energy systems are supposed to be less likely to be confronted with disruptions in low-carbon pathways and follow other studies' findings on this issue (Costantini et al., 2007; Criqui and Mima, 2012; Shukla and Dhar, 2011).

Resilience, which is generally considered in terms of the diversity of energy options in the energy system also increases as the sources of energy measured by the Shannon–Wiener diversity index (SWDI; see Elwood Shannon and Weaver, 1963) become more diverse.[12] In this context, the GEA analysis concludes that overall energy systems are supposed to be less likely to be confronted with disruptions in low-carbon pathways and follow other studies' findings on this issue (Costantini et al., 2007; Criqui and Mima, 2012; Shukla and Dhar, 2011).

---

11. The vital energy systems refer to primary energy sources and different energy carriers (electricity, hydrogen, liquid and synthetic fuels). They can be assessed at different levels as well: the end-use sectors or the TPES.

12. The index is calculated as follows:

$SWDI = -\Sigma(pi * \ln(pi))$ where pi is the share of primary energy i in total primary energy supply. In the GEA pathways, the global SWDI rises (supply diversification increases) from the current level of 1.6–2.0 by 2050, before falling to between 1.3 and 1.6 in the latter half of the century.

Moreover the reliance on a few exporting countries declines in all GEA scenarios since the energy mix is more diverse. OECD countries' energy systems, for instance, move away from fossil fuels, increase in efficiency and experience diversification in transport technologies. Energy security in China and India (which are included in a broader group) is particularly sensitive to the strong energy demand increase in these countries. Energy security improvement in both countries will depend on their ability to make their energy system more diverse and benefit from leapfroging process compared to the industrialized countries. Conversely, a risk of global climate mitigation efforts is that such measures may potentially curtail the export revenues of fossil energy producers, thereby decreasing their 'demand security'.

GEA's assessment framework has been used in a series of papers exploring the consequences of climate policies on energy security from a long-term perspective using the IMAGE and MESSAGE IAMs (Cherp et al., 2012; McCollum et al., 2013; Jewell et al., 2013, 2014). They differ in the way they explore the different possible futures. At the European level, Jewell et al. (2013) show that climate policies decrease the net energy balance after 2040 and increase the diversity of energy options for EU, particularly in the transport sector. It is also worth noting the works conducted by Guivarch et al. (2015) based on a multi-dimension analysis methodology used to capture the energy security concept, and a database of scenarios to explore the uncertainty space. Findings reveal that there is no unequivocal effect of climate policy on all the perspectives of energy security. Some indicators are improved while others are degraded, particularly in respect to the sovereignty aspects. Tensions could appear in the coming decade that are crucial in terms of implementing ambitious climate policies and could affect their acceptability. These works have to take account of the limits underlined in Gracceva and Zenewski (2014, 2015). These tensions also relate to the security of energy systems as analyzed in depth in Gracceva and Zeniewski (2014, 2015). Rising interests in the analysis of co-benefits found in the modelling agenda reflect the changes observed in the international negotiation process that pave the way for the implementation of multiple objectives policies.

7.3.3 Co-benefits and Climate Negotiations: Some Perspectives

7.3.3.1 A Potential Hook for Multiobjective Policies

The question of co-benefits has so far been poorly tackled in climate negotiations. This is all the more paradoxical that energy security concerns were one of the main motivations of the 1990 G7 meeting held in Houston, with the objective of putting a 'climate convention' on the international agenda (Kirton, 2007). Climate change was expected to be used as an argument to convince the American public to accept needed efforts to reduce oil dependence (Schlesinger, 1989). Interests in energy security has recently witnessed a revival, triggered by high oil prices at the end of the 2000's and geopolitical tensions around Russian gas but also linked to the difficulties in establishing an international climate regime that will go beyond the Kyoto Protocol.

Indeed, the focus of climate policies was primarily on adopting emission reduction objectives, under the Kyoto Protocol framework in its first stage, and more recently, through the institutionalisation of the 2°C target. The carbon market (Cap and Trade) was conceived as the main instrument for emissions reductions set at the international level. This climate-centric approach to the negotiation process that considers GHG emissions reductions as the main priority on top of development challenges showed its limits during the Copenhagen conference in 2009. It was confronted with emerging countries' unwillingness to be part of a system based on legally binding climate objectives that would potentially limit their potential growth. They argued against this approach, principally under the pretext of the Common but Differentiated Responsibility (CBDR) principle mentioned in the Climate Change Convention of 1992 (Hourcade and Shukla, 2013). Since 2009, a BU approach to climate negotiations has emerged that opens windows of opportunity to better align climate change issues with development challenges (Bodansky and Diringer, 2014; Damian, 2014; Gupta, 2014; Hourcade et al., 2015). In view of the Paris Conference in 2015 all parties to the United Nations Framework Convention on Climate Change (UNFCCC) process, including developing ones, provided their national contributions (UNFCCC, 2015). Those contributions, the so-called Intended Nationally Determined Contributions (INDCs), renamed Nationally Determined Contributions in the Paris Agreement (NDCs) are supposed to be consistent with countries' own development priorities, building upon the Nationally Appropriate Mitigation Actions (NAMAs)[13] whose framework has been set within the Copenhagen (2009) and Cancún (2010) conferences. This approach finds a particular interest among emerging countries; some of which have implemented national programmes against climate change, built around their own development objectives. This is notably the case for the Indian National Action Plan on Climate Change (NAPCC) (Dubash, 2013; Dubash et al., 2013).

---

13. In the language of UN climate negotiations the notion of NAMAs is quite wide and ambiguous (Tyler et al., 2013). It encompasses a wide array of unilateral or internationally supported measures aiming at, eg, (1) removing institutional and market failures which prevent tapping into technological potentials including those at 'negative costs'; (2) internalising the domestic co-benefits of low-carbon technologies in terms of local environment; (3) scaling up the human capacity building and technical skills, including public support for research and development to accelerate technical change beyond the pace that can be reached through the incentive of carbon prices; (4) redirecting infrastructure policies; and (5) lowering the wedges among technical, macroeconomic and social costs through appropriate macroeconomic (fiscal and financial) policies. While targeting efficient decarbonisation, the NAMAs' primary aim is to fulfil national development objectives and to do so through low-carbon instead of high-carbon development (Hourcade and Shukla, 2013). Bali Action Plan, https://unfccc.int/key_steps/bali_road_map/items/6072.php.

On the political side, there is then a growing interest in co-benefits with other policies (air quality, health, employment, energy security, biodiversity protection) in order to alleviate the negative impacts of climate policies such as increased energy prices for poor households. The EU is involved in such an approach: both the Climate and Energy Package and the Energy Roadmap 2050 (European Commission, 2011) endorse the goals of reducing GHG emissions while ensuring the security of energy supplies. China's ambitious objectives in terms of energy efficiency occur alongside strong efforts to increase energy resources access throughout the world (Sovacool and Brown, 2010) whilst the US applies more stringent fuel consumption standards for cars (Anderson et al., 2011), in order to reduce their oil dependency by one-third by 2025 (The White House, 2011). Indeed, most policies that impact emissions reduction objectives are motivated by reasons that are not specifically related to climate but linked to other issues including energy security and urban air pollution.

7.3.3.2 Challenges of Representing Multiple Objectives in Integrated Assessment Models

The example of energy policies shows how 'non-climate policies' determine a significant share of GHG emissions and represent essential levers to comply with ambitious mitigation objectives (Hourcade and Shukla, 2013). Modelling studies of Indian Institute of Management Ahmedabad (IIMA) or from CIRED using the IMACLIM model show that the implementation of complementary policies to a carbon price, for instance in the transport sector, tends to reduce the net cost of climate policies, particularly in developing countries (Shukla and Dhar, 2011; Waisman et al., 2012; Hamdi-Cherif and Waisman, 2015). These results are similar to the conclusions of the GEA regarding the importance of conducting multiobjective policies that make poverty eradication, energy access for billions of people in southern countries and the pursuit of ambitious emissions reduction objectives possible in the coming decades.

The implementation of multiobjective policies therefore leads to a broader scope in terms of climate policy instruments. A climate centric approach tends indeed to favour specific types of policy instruments such as carbon taxes or carbon markets while other instruments designed to better integrate climate policies within non-climate policies are underestimated (Lecocq, 2015). Flagship policies such as the EU Emissions Trading System (ETS) should not hide the fact that the fight against climate change is often considered as an objective among others, even as a co-benefit of development policies. Analysing the co-benefits of development policies on emissions can lead to a focus on a set of mechanisms that drive emissions such as those related to land dynamics (particularly taxation) or to the financing of the economy. These mechanisms have more or less direct impacts on the dynamics of energy systems and finally on the three main perspectives identified in energy security (sovereignty, resilience and robustness). A better integration of these mechanisms in IAMs represents modelling frontiers for the coming years.

7.4 CONCLUSION

The scenarios analysed in the fifth IPCC report show that achieving the 2°C objective is feasible (1) if countries commit themselves to reaching this goal as early as possible as part of a global international agreement and (2) if there is a rapid deployment of key low-carbon technologies. Although climate change and energy security were originally two interlinked issues, this chapter highlighted that only a few scenarios in the IPCC database of Group III address the potential synergies between climate change and energy security issues. This is partly the result of the agenda of international negotiations being overly centered on climate emission reduction targets for the last 15 years and less on linking the agenda of climate policies with other development issues, in particular energy security. Climate talks have gradually come to be structured around the objective of limiting global temperature rise to 2°C above preindustrial levels, which have then, in turn, impacted research issues in modelling exercises. Recent changes in climate negotiations have, however, paved the way for a better alignment of climate policies with development issues. The GEA works have provided interesting quantitative highlights on the potential co-benefits and trade-offs of stringent climate policies. However, their quantification is still an ongoing research question.

Key methodological issues remain regarding the representation of other determinants of emissions that relate to specificities of urban planning, infrastructures, the nature of the labour market, consumers' behaviours and societal processes, etc.[14] These factors can also impact the way societies consume energy and the resilience of energy systems to internal and external disruptions. These challenges provide the opportunity to develop better models to elaborate more comprehensive ways to address the multifaceted nature of energy security issues (co-benefits/adverse side effects) in a low-carbon context.

Indeed, the debate about the effect of climate change policies on energy security strategies may defy resolution precisely due to the methodologies that are typically applied to both sides of the argument. Often a handful of parameters to represent 'energy security' and 'climate change' are selected for comparison and analysis under a set of different long-term energy scenarios. Our view is that a critical issue to be addressed is the need for a disciplinary pivot away from indicator-based assessments, given their partial and simplified view of energy security. On the one hand, energy security is a product of many diverse attributes, from the diversity of gas imports to the capacity margins in the power sector; therefore, it needs to be assessed from a systemic perspective that takes all of these attributes into account (Watson, 2009). On the other hand, energy security is a product of the interactions and interdependencies of a complex

---

14. By societal processes, we mean a range of grass-roots initiatives and experiences of low energy and carbon transition at a local level. Some pave the ground for decentralised energy production modes.

system, one 'whose properties are not fully explained by an understanding of its component parts' (Gallagher and Appenzeller, 1999).

The rest of Section IV of this book will address this issue by first discussing the intrinsic complexity of energy security (Chapter 9), and then proposing a conceptual and methodological approach to energy security that is consistent with its polysemic, multidimensional and 'systemic' nature (Chapter 10).

ACKNOWLEDGEMENT

The authors acknowledge the EUFP7 Milesecure-2050 project for valuable research support and useful comments from Céline Guivarch.

REFERENCES

Amerighi, O., Ciorba, U., Tommasino, M.-C., 2010. Models Characterization Report. ATEST Project EUFP7.

Anderson, S.T., Parry, I.W., James, H., Sallee, M., Fischer, C., 2011. Automobile fuel economy standards: impacts, efficiency, and alternatives. Review of Environmental Economic and Policy 5 (1), 89–108.

Augutis, J., Krikstolaitis, R., Martisauskas, L., Peciulyte, S., 2012. Energy security level assessment technology. Applied Energy 97, 143–149.

Babonneau, F., Kanudia, A., Labriet, M., Loulou, R., Vial, J.-P., 2009. Energy security: a robust programming approach and application to European energy supply via TIAM. In: PLANETS (Probabilistic Long-term Assessment of New Energy Technology Scenarios) Project, DELIVERABLE No 15, Report on Probabilistic Scenarios, Chapter 9.

Baritaud, M., 2012. Securing Power During the Transition. Generation Investment and Operation Issues in Electricity Markets with Low-Carbon Policies. IEA Insights paper, OECD/IEA, Paris.

Benson, S.M., 2014. Negative-emissions insurance. Science 344 (6191), 1431.

Bibas, R., Mejean, A., 2014. Potential and limitations of bioenergy options for low carbon transitions. Climatic Change 123 (3), 731–761.

Bibas, R., 2015. Methodological, Technical and Macroeconomic Insights on the Energy Transition: Forward-Looking Analysis, Technologies and Investment (Ph.D. thesis). ENPC-CIRED. 526 pp.

Bodansky, D., Diringer, E., 2014. Evolution of the International Climate Effort. Center for Climate and Energy Solutions, Arlington. 20 pp.

Brown, S.P.A., Huntington, H.G., 2008. Energy security and climate change protection: complementarity or trade-off? Energy Policy 36 (9), 3510–3513.

Brunelle, T., 2012. The Impact of Global Drivers on Agriculture and Land-Use (Ph.D. thesis). AgroParisTech-CIRED. 159 pp.

Cassen, C., Guivarch, C., Lecocq, F., 2015. Les cobénéfices des politiques climatiques : un concept opérant pour les négociations climat? Natures Sciences Sociétés 23 (Suppl.), 41–51.

Chaudry, M., Ekins, P., Ramachandran, K., Shakoor, A., Skea, J., Strbac, G., Wang, X., Whitaker, J., 2009. Building a Resilient UK Energy System. Working Paper, vol. 23. United Kingdom Economic Research Council, London.

Cherp, A., Jewell, J., 2011. Measuring energy security: from universal indicators to contextualised frameworks. In: Sovacool, B.K. (Ed.), The Routledge Handbook of Energy Security. Routledge, Abingdon, New York, p. 1818.

Cherp, A., Adenikinju, A., Goldthau, A., Hernandez, F., Hughes, L., Jansen, J., Jewell, J., Olshanskaya, M., de Oliveira Soares, R., Sovacool, B., Vakulenko, S., 2012. Energy and security. In: Global Energy Assessment: Toward a More Sustainable Future. Cambridge University Press and Laxenburg, Cambridge, UK and New York, NY, USA, pp. 325–383 (Austria: The International Institute for Applied Systems Analysis, Cambridge University Press).

Chester, L., 2010. Conceptualising energy security and making explicit its polysemic nature. Energy Policy 38, 887–895.

Clarke, L., Edmonds, J., Krey, V., Richels, R., Rose, S., Tavoni, M., 2009. International climate policy architectures: overview of the EMF 22 international scenarios. Energy Economics 31, S64–S81.

Clarke, L., Jiang, K., Akimoto, K., Babiker, M., Blanford, G., Fisher-Vanden, K., Hourcade, J.-C., Krey, V., Kriegler, E., Löschel, A., McCollum, D., Paltsev, S., Rose, S., Shukla, P.R., Tavoni, M., van der Zwaan, B.C.C., van Vuuren, D.P., 2014. Assessing transformation pathways. In: Edenhofer, O., Pichs-Madruga, R., Sokona, Y., Farahani, E., Kadner, S., Seyboth, K., Adler, A., Baum, I., Brunner, S., Eickemeier, P., Kriemann, B., Savolainen, J., Schlömer, S., von Stechow, C., Zwickel, T., Minx, J.C. (Eds.), Climate Change 2014 Mitigation of Climate Change: Contribution of Working Group III to the Fifth Assessment Report of the Intergovernmental Panel on Climate Change. Cambridge University Press, Cambridge (UK) and New York.

Costantini, V., Gracceva, F., Markandya, A., Vicini, G., 2007. Security of energy supply: comparing scenarios from a European perspective. Energy Policy 35, 210–226. http://dx.doi.org/10.1016/j.enpol.2005.11.002. Available at: http://www.sciencedirect.com/science/article/pii/S0301421505003009.

Crassous, R., 2008. Modéliser le long terme dans un monde de second rang: application aux politiques climatiques (Ph.D. thesis). AgroParisTech.

Criqui, P., Mima, S., 2011. The European Energy Sector and the Climate-Security Nexus in the SECURE Scenarios. Background Papers. Gulf Research Foundation.

Criqui, P., Mima, S., 2012. European climate—energy security nexus: a model based scenario analysis. Modeling Transport (Energy) Demand and Policies 41, 827–842. http://dx.doi.org/10.1016/j.enpol.2011.11.061. Available at: http://www.sciencedirect.com/science/article/pii/S0301421511009591.

Damian, M., 2014. La politique climatique change enfin de paradigme. Économie Appliquée 17 (1), 37–72.

DBERR, 2007. Gas Supply Shock Analysis. Department for Business Enterprise and Regulatory Reform, London. 262 pp.

Dubash, N.K., Raghunandan, D., Sant, G., Sreenivas, A., 2013. Indian climate change policy: exploring a co-benefits based approach. Economic and Political Weekly 48 (22), 47–61.

Dubash, N.K., 2013. The politics of climate change in India: narratives of equity and cobenefits. Wiley Interdisciplinary Review of Climate Change 4 (3), 191–201.

Elwood Shannon, C., Weaver, W., 1963. The Mathematical Theory of Communication. Univ. of Illinois Press.

European Commission, 2006. Green Paper on a European Strategy for Sustainable. Competitive and Secure Energy. Brussels.

European Commission, 2010. Energy 2020 a Strategy for Competitive, Sustainable and Secure Energy. Brussels.

European Commission (EC), 2011. Energy Roadmap 2050. COM 2011, 885 final.

European Commission, 2013. Green Paper on a 2030 Framework for Climate and Energy Policies. Brussels.

EUROSTAT, 2015. EUROSTAT statistics explained.

Fuss, S., Canadell, J.G., Peters, G.P., Tavoni, M., Andrew, R.M., Ciais, P., Jackson, R.B., Jones, C.-D., Kraxner, F., Nakicenovic, N., Le Quéré, C., Raupach, M.R., Sharifi, A., Smith, P., Yamagata, Y., 2014. Betting on negative emissions. Nature Climate Change 4 (10), 850.

Gallagher, R., Appenzeller, T., 1999. Beyond reductionism. Science 284 (5411), 79.

Gracceva, F., Zeniewski, P., 2014. A systemic approach to assessing energy security in a low-carbon EU energy system. Applied Energy 123, 335–348.

Gracceva, F., Zeniewski, P., 2015. Between climate change and energy security objectives: the potential role of shale gas. In: Chapter of the Handbook for Clean Energy Systems. Wiley.

Greenleaf, J.R.H., Angelini, T., Green, D., Williams, A., Rix, O., Lefevre, N., 2009. Analysis of Impacts of Climate Change Policies on Energy Security. Ecofys. 352 pp.

Guivarch, C., Monjon, S., Rozenberg, J., Voght-Shilb, A., 2015. Would climate policy improve the European energy security? Climate Change Economics 6 (2), 1550008-35 pp.

Guivarch, C., 2009. Evaluating Mitigation Costs, the Importance of Representing Second Best Mechanisms (Ph.D. thesis). Ecole des Ponts ParisTech, Paris Est University.

Gupta, J., 2014. The History of Global Climate Governance. Cambridge University Press, Cambridge, UK. 264 pp.

Haas, R., Lettner, G., Auer, H., Duic, N., 2013. The looming revolution: how photovoltaics will change electricity markets in Europe fundamentally. Energy 57 (0), 38–43.

Haberl, H., Beringer, T., Bhattacharya, S.C., Erb, K.H., Hoogwijk, M., 2010. The global technical potential of bio-energy in 2050 considering sustainability constraints. Current Opinion on Environmental Sustainability 2 (5–6), 394–403.

Hamdi-Cherif, M., Waisman, H.-D., 2015. Global carbon pricing and the "common but differentiated responsibilities" – the case of China. online version. In: International Environmental Agreements: Politics, Law and Economics, pp. 1–19. http://dx.doi.org/10.1007/s10784-015-9289-2.

Hourcade, J.-C., Shukla, P.R., 2013. Triggering the low-carbon transition in the aftermath of the global financial crisis. Climate Policy 13 (Suppl. 1), 22–35 Low carbon drivers for a sustainable world.

Hourcade, J.C., Jaccard, M., Bataille, C., Ghersi, F., 2006. Hybrid modeling: new answers to old challenges. The Energy Journal (Special issue 2), 1–11.

Hourcade, J.-C., Shukla, P.-R., Cassen, C., 2015. Climate policy architecture for the Cancun paradigm shift: building on the lessons from history. International Environmental Agreements: Politics, Law and Economics 15 (4), 353–367.

Hughes, L., 2009. The four 'R' of energy security. Energy Policy 37, 2459–2461.

IIASA, 2012. Global Energy Assessment (GEA): Toward a Sustainable Future. Cambridge University Press, Cambridge, UK.

IPCC, 2007. Mitigation of Climate Change: Contribution of Working Group III to the Fourth Assessment Report of the Intergovernmental Panel on Climate Change. Cambridge University Press, New York.

IPCC, 2011. IPCC special report on renewable energy sources and climate change mitigation. In: Edenhofer, O., Pichs-Madruga, R., Sokona, Y., Seyboth, K., Matschoss, P., Kadner, S., Zwickel, T., Eickemeier, P., Hansen, G., Schlömer, S., von Stechow, C. (Eds.), Prepared by Working Group III of the Intergovernmental Panel on Climate Change. Cambridge University Press, Cambridge, UK.

IPCC, 2014. In: Edenhofer, O., Pichs-Madruga, R., Sokona, Y., Farahani, E., Kadner, S., Seyboth, K., Adler, A., Baum, I., Brunner, S., Eickemeier, P., Kriemann, B., Savolainen, J., Schlömer, S., von Stechow, C., Zwickel, T., Minx, J.C. (Eds.), Climate Change 2014: Mitigation of Climate Change: Contribution of Working Group III to the Fifth Assessment Report of the Intergovernmental Panel on Climate Change. Cambridge University Press, Cambridge (UK) and New York.

Jansen, J.C., Arkel, W.G., Boots, M.G., 2004. Designing Indicators of Long-term Energy Supply Security. Energy Research Centre Netherlands (ECN), Petten.

Jewell, J., Cherp, A., Vinichenko, V., Bauer, N., Kober, T., McCollum, D.L., van Vuuren, D.P., van der Zwaan, B., 2013. Energy security of China, India, the E.U. and the U.S. under long-term scenarios: results from six IAMs. Climate Change Economics 4 (4), 1340011.

Jewell, J., Cherp, A., Riahi, K., 2014. Energy security under de-carbonization scenarios: an assessment framework and evaluation under different technology and policy choices. Energy Policy 65, 743–760.

Jewell, J., 2011. Model of Short-term Energy Security (MOSES). IEA, Paris.

Jun, E., Kim, W., Chang, S.H., 2009. The analysis of security cost for different energy sources. Applied Energy 86 (10), 1894–1901.

Kirton, J., 2007. The G8's energy-climate connection. In: Workshop or Talking Shop? Globalization, Security and the Legitimacy of the G8, Brussels, 24th May.

Kruyt, B., van Vuuren, D.P., de Vries, H.J.M., Groenenberg, H., 2009. Indicators for energy security. Energy Policy 37 (6), 2166–2181.

Labriet, M., Loulou, R., Kanudia, A., 2009. Modeling uncertainty in a large scale integrated energy-climate model. In: Filar, J.A., Haurie, A. (Eds.), Uncertainty and Environmental Decision Making: A Handbook of Research and Best Practice. International Series in Operations Research and Management Science, 138. Springers, pp. 51–77.

Lambin, E., Meyfroidt, P., 2011. Global land use change, economic globalization, and the looming land scarcity. Proceedings of the National Academy of Sciences of the United States of America 108 (9), 3465–3472.

Lecocq, F., 2015. L'évaluation des politiques climatiques à l'échelle globale et nationale. In: Torquebiau, E. (Ed.), Changer d'habitudes: changement climatique et agriculture du monde. Quae, Versailles, pp. 279–291.

Lefèvre, N., 2007. Energy Security and Climate Policy. IEA, Paris.

Lefèvre, N., 2010. Measuring the energy security implications of fossil fuel resource concentration. Energy Policy 38 (4), 1635–1644.

Loschel, A., Moslener, U., Rubbelke, D., 2010. Energy security-concepts and indicators. Energy Policy 38 (Editorial Issue), 1607–2074.

Luft, G., Korin, A., Gupta, E., 2011. Energy security and climate change: a tenuous link. In: Sovacool, B.K. (Ed.), The Routledge Handbook of Energy Security. Routledge, Abingdon (England), New York.

Maisonnave, H., Pycroft, J., Saveyn, B., Ciscar, J.-C., 2012. Does climate policy make the EU economy more resilient to oil price rises? A CGE analysis. Energy Policy 47, 172–179.

Markandya, A., Halsnaes, K., Lanza, A., Matsuoka, Y., Maya, S., Pan, J., Shogren, J., Seroa de Motta, R., Zhang, T., 2001. In: Metz, B., Davidson, O., Swart, R., Pan, J. (Eds.), Costing Methodologies, Contribution of Working Group III to the Third Assessment Report of the Intergovernmental Panel on Climate Change. Cambridge University Press, UK. 700 pp.

Martchamadol, J., Kumar, S., 2013. An aggregated energy security performance indicator. Applied Energy 103, 653–670.

McCollum, D.L., Krey, V., Riahi, K., Kolp, P., Grubler, A., 2013. Climate policies can help resolve energy security and air pollution challenges. Climatic Change 119 (2), 479–494.

Meadows, D.H., Meadows, D.L., Randers, J., Behrens, W.W., 1972. The Limits to Growth. Universe Books, New York, NY, USA. 1075 pp.

van der Meer, J.W.M., Huppert, H., Holmes, J., 2014. Carbon: no silver bullet. Science 345 (6201), 1130.

Riahi, K., Dentener, F., Gielen, D., Grubler, A., Jewell, J., Klimont, Z., Krey, V., McCollum, D., Pachauri, S., Rao, S., van Ruijven, B., van Vuuren, D.P., Wilson, C., 2012. Chapter 17-Energy Pathways for Sustainable Development. In: Global Energy Assessment - Toward a Sustainable Future. Cambridge University Press, Cambridge, UK and New York, NY, pp. 1203–1306. USA and the International Institute for Applied Systems Analysis, Laxenburg, Austria.

Rozenberg, J., Hallegatte, S., Vogt-Schilb, A., Sassi, O., Guivarch, C., Waisman, H., 2010. Climate policies as a hedge against the uncertainty on future oil supply. Climatic Change 101 (3–4), 663–668.

Sarofim, M., Reilly, J., 2011. Applications of integrated assessment modeling to climate change. Wiley Interdisciplinary Review: Climate Change 2 (1), 27–44.

Schlesinger, R., 1989. Energy and geopolitics in the 21st century. In: Communication at the 14th World Energy Congress, Montreal, Canada.

Shukla, P.R., Dhar, S., 2011. Climate agreements and India: aligning options and opportunities on a new track. International Environmental Agreements: Politics, Law and Economics 11 (3), 229–243.

Shuttleworth, G., 2002. Security in Gas and Electricity Markets, Final Report for the Department of Trade and Industry. NERA, London.

Skea, J., Ekins, P., Winskel, M. (Eds.), 2011. Energy 2050. Making the Transition to a Secure and Low-Carbon Energy System. Earthscan.

Souty, F., 2013. Modélisation de l'évolution des surfaces agricoles à l'échelle de grandes régions du monde (Ph.D. thesis). AgroParisTech. 186 pp.

Sovacool, B.K., Brown, M.A., 2010. Competing dimensions of energy security: an international perspective. Annual Review of Environment and Resources 35, 77–108.

Stirling, A., 2009. The Dynamics of Security. Stability, Durability, Resilience, Robustness Energy Security in a Multipolar World. Kohn Centre, Royal Society, London.

Stirling, A., 2011. The diversification dimension of energy security. In: Sovacool, B.K. (Ed.), The Routledge Handbook of Energy Security. Routledge.

The White House, 2011. Blueprint for a Secure Energy Future. Washington DC, online: http://www.whitehouse.gov/sites/default/files/blueprint_secure_energy_future.pdf.

Tilman, D., Clark, M., 2014. Global diets link environmental sustainability and human health. Nature 515 (7528), 518–522.

Tyler, E., Boyd, A., Coetzee, K., Torres Gunfaus, M., Winkler, H., 2013. Developing country perspectives on 'mitigation actions', 'NAMAs', and 'LCDS'. Climate Policy 13 (6), 770–776. http://dx.doi.org/10.1080/14693062.2013.823334.

UNFCCC, 2015. Lima Call for Climate Action. Decision 1/CP.20, online: http://unfccc.int/resource/docs/2014/cop20/eng/10a01.pdf.

Ürge-Vorsatz, D., Tirado Herrero, S., Dubash, N.K., Lecocq, F., 2014. Measuring the co-benefits of climate change mitigation. Annual Review of Environment and Resources 39 (1), 549–582.

Van Vuuren, D.P., Bellevrat, E., Kitous, A., Issac, M., 2010. Bio-energy use and low stabilization scenarios. The Energy Journal 31 (Special issue 1), 192–222.

Waisman, H., Guivarch, C., Grazi, F., Hourcade, J.-C., 2012. The Imaclim-R model: infrastructures, technical inertia and the costs of low carbon futures under imperfect foresight. Climatic Change 114 (1), 101–120.

Watson, J., 2009. Is the move toward energy security at odds with a low-carbon society? In: Giddens, A., Latham, S., Liddle, R. (Eds.), Building a Low-Carbon Future. The Politics of Climate Change. Policy Network, London.

Weyant, J., Kriegler, E., Blanford, G., Krey, V., Edmonds, J., Riahi, K., Richels, R., Tavoni, M. (Eds.), 2014. Special Issue: The EMF27 Study on Global Technology and Climate Policy Strategies, 123, pp. 3–4.

Winzer, C., 2012. Conceptualizing energy security. Energy Policy 46, 36–48.

Wu, G., Liu, L.-C., Han, Z.-Y., Wei, Y.-M., 2012. Climate protection and China's energy security: win–win or tradeoff. Applied Energy 97, 157–163.

Zhang, Z., Folmer, H., 1998. Economic modeling approaches to cost estimates for the control of carbon dioxide emissions. Energy Economics 20 (1), 101–120.

Chapter 8

Towards Governance of Energy Security

G. Valkenburg[1], F. Gracceva[2]

[1]*Maastricht University, Maastricht, The Netherlands;* [2]*Studies and Strategies Unit, ENEA (Italian National Agency for New Technologies, Energy and Sustainable Economic Development), Rome, Italy*

8.1 INTRODUCTION: TRANSITION AS AN UNSTRUCTURED PROBLEM

In order to secure low-carbon energy provision for the future, transitions in the energy system are needed: a great number of incremental as well as fundamental changes, together engendering a radical overhaul of the energy system. Energy transitions involve not only technological changes but also changes in institutions, social structures and market positions (Lockwood et al., 2013; Miller et al., 2013). The ramifications of these changes and their interdependencies engender a proliferation of complexity. This makes managing transitions an elusive affair: changes in one realm may cause the consequences of interventions in another realm to become less predictable or less effective. We argue in this chapter that such complexity is not simply resolved by conducting further research or by bringing in new knowledge. This chapter exposes some complexities that typically occur when energy transitions are initiated in pursuit of energy security. Consequently, it reviews what policy studies have argued about the governance of complex problems in general. Finally, some strategies will be proposed regarding the management of energy transitions.

For any problem to be solved, there must be sufficient alignment between the problem and the available tools or solutions. However, as transitions are highly heterogeneous affairs, problems interfere, conflate and become multifaceted (Stirling, 2014): social issues render technological aspects ambiguous, technological puzzles drive social issues to a head, different sectors with different preferences for solutions have to cooperate and problems as well as solutions run across all levels, from individual life to global structures. It thus becomes highly unclear which tools are to be used. Conventionally, one might be tempted to think that a mismatch is to be resolved by revision or adaptation of the tools. Yet with problems of the complex kind under discussion here, it is

Low-carbon Energy Security from a European Perspective. http://dx.doi.org/10.1016/B978-0-12-802970-1.00008-5
Copyright © 2016 Elsevier Ltd. All rights reserved.

increasingly the case that no appropriate tools are available and that the problem itself requires modification: something must be sufficiently similar to a nail for a hammer to be the tool of choice.

The mismatch between problems and solutions is in this case the consequence of the heterogeneity of the problems. For example, the change from centralised coal-fired power plants to the use of photovoltaic installations on private roofs is not only a technical challenge of rebalancing the flows of electrical energy through the power grid. Moving to solar power also entails revision of institutional, economic and social structures, such as the way energy markets are divided and whose interests are prioritised in the management of the energy network. This ties in with a large number of different stakeholders, different forms of expertise and sorts of knowledge in general, and different sites of (sociopolitical) power and agency that bear on the problem and its solutions. Altogether, it means that there is no self-evident approach to the implementation of solar power.

More generally, the possibility of an energy transition is normally assessed by looking at its potential impact in terms of the classical three corners of the energy policy dilemma (see, for instance, European Commission, 2006): energy security, carbon reduction and competitiveness. However, the exact specification of these values is controversial, connected to ideology and subject to negotiation. Also, their specification depends on contextual particularities, on factual knowledge that is largely unavailable, and on the timeframe we take into consideration. The question of what kind of low-carbon and energy-secure world we want to build in fact contains all these sub questions, calling for a host of different, incompatible and mutually detrimental solutions.

Because of these complexities, transitions towards secure low-carbon systems qualify as unstructured problems: it is literally unclear from the problem itself what the tool of choice should be or how the problem should be approached at all. Conklin (2005) points out a number of difficulties that typically emerge with such unstructured problems. For one thing, it is impossible to fully understand the problem prior to solving it. The problem becomes clear first in the course of hammering out a solution. Also, there is no clear definition of what makes a solution successful or final, and one can only try one solution rather than many. The problem itself is unique as well, which limits the possibility to apply lessons learnt from earlier problems.

In the case of low-carbon energy security, such lack of structure already follows from the example just given. For one thing, technological changes will have to be matched with the aforementioned social changes. Moreover, the 'stuff' on which energy transition is to operate is highly heterogeneous: not only copper wires and institutions matter, also markets, geopolitics, human behaviour and societal aspects as well as natural disasters and their abatement are ultimately within the sphere of what matters to energy transitions. Notably, there is no priority between these different realms, as no realm has an essential primacy over another. This means that social, technical, economic, political and any other change will have to be made in concert. This adds to the ambiguity in the choice of methods, both for production of knowledge and for the implementation of new sociotechnical configurations.

This stands in stark contrast to the idea that governance often seems to be concerned with what we could call propositional questions: questions that can be approached by clear inferences that lead to an unambiguous conclusion. As Jasanoff (2003, 2007) observes, policy problems are typically discussed in a frame that seems to uncritically adopt the criteria that are also believed to determine what 'good natural science' is. More specifically, this means that the values that typically guide natural-scientific research are taken to be the criteria for policy making as well: transparency, reproducibility, impersonality, objectivity, etc. A prime example is the translation of phenomena into simple mathematical relations like quantitative indicators. However, because of their lack of structure, complex problems in general and low-carbon energy security in particular resist such 'scientistic' framing. By consequence, they resist the solutions that come about from such framings.

Not only do complex problems resist reduction into simple quantitative indicators, they also cannot easily be made less complex. Their complexity cannot be reduced by enrolling additional expertise or by other partial solutions. Nor is it a matter of conducting additional political debate so as to acquire better knowledge of what a society wants or should do. As will be shown onwards the complexity of the problems is not the consequence of a knowledge deficit, or at least a clearly defined knowledge deficit for which indeed an expert could be consulted or a partial solution could be mobilised. Rather, the complexity is innate to the problem itself.

This chapter discusses how the fundamental ambiguity of energy security is an important factor to explain the complexity of the governance of energy transition. The discussion fleshes out in more detail how these ambiguities can and cannot be resolved so as to forge the problem into a manageable issue. By addressing specific dimensions of the underlying complexities, partial reductions of complexity can be achieved, thus bringing issues closer to the propositional issues that policy can deal with. This step proves necessary for governance of energy transitions and energy security. This chapter is based on theoretical and conceptual research, drawing on literature from social studies of science and technology, expertise studies and studies of the public understanding of science, and notably economics and its perspectives on energy security.

This chapter is organised by the following logic. In the next section, it will be discussed how energy security is normatively problematic: it remains highly disputable how the value of energy security should ultimately be realised. Then, in the third section, it will be shown how energy security is fundamentally difficult to measure or otherwise assess in an empirical sense. From the fourth section onwards, it will be further developed how such problems, subject to both normative and empirical uncertainty, can be approached and be made subject to governance.

8.2 THE CONCEPTUAL BABYLON OF ENERGY SECURITY

There is no such thing as energy security. Rather, there are many different securities, or different conceptions of the value that compete for relevance. In this section, we provide an overview of the conceptual intricacies that the concept of energy security hides. Together with the empirical difficulties provided in the next section, this sketches the landscape in which governance of energy transitions is to operate.

It has been frequently claimed that it is difficult even to converge on a definition of energy security (eg, Sovacool and Mukherjee, 2011). Indeed a key problem with energy security is that there is no common understanding on which values are relevant and on how they should be prioritised.

A reason for the normative ambiguity of energy security stems from the fact that this concept is largely defined at the level of society: the population at large should have a certain degree of access to energy, and particular institutions are entitled to even a higher degree of access certainty because of the societally disrupting effect their shutting down would have. The exact distribution of the burden of maintaining this security as well as the burden of the costs of an eventual shutdown remains a matter of social justice, and ideologically informed at that. A debate concerning such social justice issues would be heavily hampered by lacking factual knowledge about the exact dangers that are to be coped with: the 'risks' that form the input of decision-making processes are themselves only proxies of perceived (even if reasonable and realistic) threats. We will discuss the problem of such partial knowledge in the following section, but it is already visible here that it is not quite bringing the problem any closer to a solution.

A further reason for the normative ambiguity of energy security stems from its multidimensional and polysemic nature: it harbours a range of different risks of geological, technical, economic, geopolitical, social and environmental nature (Checchi et al., 2009). Clearly, these dimensions are not limited to the functioning of energy markets. For instance, environmental aspects and a concern for sustainable development are often included in the concept of energy security (European Commission, 2000; Asia Pacific Energy Research Centre, 2007; United Nations Development Programme, 2004). Not only are there different dimensions and aspects to energy security, but also is it the case that different perspectives will typically attribute different relative importance to those aspects. Moreover, for this relative importance it matters also whether a short- or long-term perspective is adopted: depending on the timeframe under consideration the analysis will include or not issues such as short-term operational maintenance, efficient investments in the medium and long-term, or adequate regulation for the whole value-chain, from resource extraction and/or import to end use. This is well reflected in the way energy security has evolved with the transformation of the world's energy regime (Chester, 2010).

The differences in relative importance attributed to such dimensions are brought out in the following discussion by elaborating two received perspectives on energy security: an economic perspective and a strategic geopolitical one. We deliberately choose to take two perspectives from different disciplines as opposed to splitting out the details of different frameworks from a single discipline.

These two particular disciplines are chosen because they produce outlooks that are fundamentally different and not reconcilable through any form of negotiation. As will prove relevant in the next section on governance of complex problems, discussing two radically different perspectives clearly shows the fundamental nature of the normative ambiguities. If we had chosen to discuss different 'schools' from within the same discipline, we would in all likelihood not have been able to completely take away the impression that remaining disagreement could still have been further reduced by additional research or debate. This all leaves unaffected, though, that both positions on their own terms are legitimate, rational, and thoughtful.

8.2.1 A Geopolitical Perspective on Energy Security

A geopolitical perspective typically stresses the political and strategic side of energy security. Since any uncertainty in the supply and price of energy can threaten the working of a well-functioning modern society, energy is recognised as a 'strategic commodity'. Consequently, energy security is today again, as it was in the 1980s, a matter of national security (Yergin, 2006). It is imperative for a country to ensure secure energy supply at affordable rates, whether through international cooperation, government intervention or military control.

One thing clearly visible in this perspective is that, with the access to energy depending on international networks of infrastructure and transport, the policies of energy-rich countries are characterised by unilateral approaches based on power politics, rather than following approaches of international law and multilateral political cooperation. Particularly outside Europe, energy policies are determined more than ever by the strategic and geopolitical interests of national foreign and security policies.

This perspective depends on the geopolitical assumption that 'states are the primary actors in the system, [that] a military-economic competition exists between them for the raw materials needed for national power, [that] states and alliances are able to "balance" one another either through physical occupation or by securing political influence within a geographical space, and that geography represents perhaps the greatest determinant of political relationships' (Mayer, 2013). Thus, this perspective gives primary importance to the interplay of natural resources, strategic dominance and geographic space on one hand, and the various state and nonstate actors on the other.

Viewing the history of the world of the 20th century through this lens shows a bleak picture of what energy security can amount to. It has prompted a wide array of atrocities, notably between the West and the Arab peoples (Maugeri, 2006). Examples range from the first oil-driven policies at the dawn of the 20th century, leading to the British control of Persia and the establishment of today's Iraq, to the close links between the American government and Saudi Arabia's and more generally the American foreign policy in the Middle East. A most striking example is the 'Carter doctrine', which reckoned the necessity of stability in the Persian Gulf region a legitimation for military force to be applied.[1] More recently, Russia's foreign policy is often taken to be centred on using energy exports as an instrument of force and even of intimidation or blackmail (Umbach, 2010).

More widely, this geopolitical approach can also be discerned behind the fear of a clash of civilisations between Islam and the rest of the world, which would threaten access to the largest oil deposits. Additionally, this geopolitical approach can be found behind the national policies followed by European Union (EU) member states within the EU energy markets: there is a certain reluctance to integrate national energy markets. It is feared that foreign energy companies benefit from national subsidies and support schemes, or from the development of national capacity mechanisms.

In contrast to the economic perspective to be discussed below, the above geopolitical perspective on energy security seemingly assumes that the market alone is not able to deal with the multifaceted challenges affecting the supply of energy, which include political conflicts and instability of export and transit countries, wars, resource nationalism, accidents and sabotage. Even if globalisation of energy markets avails various energy resources across the world, they remain strategic goods. As such, they are part of the foreign and security policy strategy of energy-rich countries, whose energy policies do not necessarily adhere to the rules of market economics (Umbach, 2010). This renders highly questionable the idea that the security of energy supply can be left to 'smoothly functioning international energy markets' (International Energy Agency, 2002, p. 3). Energy security strategies are likely to keep playing roles that go beyond overcoming 'situation when energy markets do not function properly' (Noel, 2008a,b).

8.2.2 An Economic Perspective on Energy Security

Economic perspectives identify the primary issues in energy security at the avoidance of economic loss, ie, lost production or consumption opportunities caused by threats to the energy supply chain (Kaderják, 2011; Bohi and Toman, 1996). According to this approach, energy insecurity is indeed 'the loss of welfare that may occur as a result of a change in the price or availability of energy' (Bohi and Toman, 1996). Since any energy system delivers some level of security for consumers, the real question is: 'How much security is enough?' (National Economic Research Associates, 2002), and the answer is that the 'right' level of security 'depends on the balance between the costs and the benefits of increasing security' (NERA, 2002; Department for Business, Enterprise and Regulatory Reform, 2007).

According to this view, if there were perfectly competitive and complete markets for security and risk reduction, the market outcome would be the optimal level of security. First, the economic value of energy security is in that case

---

1. President of the United States of America Jimmy Carter, *State of the Union*, 23 January 1980.

a function of the consumers' willingness to pay for any unit of added energy security. This willingness tends to fall if risks are reduced. Second, the economic value is a function of the costs of providing additional security, which tend to rise if the risk level is reduced. In practice, calculating this market outcome would require knowing both the willingness curve and the cost curve, which is arduous to achieve. However, for most kinds of security issues, it is in principle possible to reach a sub-optimal situation where security remains within a 'zone of adequacy' (NERA, 2002).[2]

Contrary to the geopolitical perspective, an economic perspective sees energy resources as exclusively economic goods, not strategic ones. Accordingly, political and hard power factors are left outside the analysis. By consequence, policy for supply security must primarily strike a balance between the costs of improving supply security and the benefits from it in the form of a reduction of welfare loss owing to nonprovided energy services (Kaderják, 2011).

In fact, the main role of energy security policies is just to focus on the 'situations when energy markets do not function properly' (Noel, 2008a), due to a number of possible reasons: market failures, the presence of barriers preventing the possibility of a market for security and risk reduction, the public good characteristics of energy security nature.[3] However, the general principle is that competitive markets and independent regulation are considered the 'most effective way of delivering secure and reliable energy supplies' (DBERR, 2007), because suppliers are better placed than government or the regulator to understand the value that their different customers place on security of supply (DBERR, 2007). Therefore, even in case of a market failure, policy interventions are justified only if it is demonstrated that its effect would improve the level of security.

Such an economic perspective is reflected in belief held by the International Energy Agency that 'smoothly functioning international energy markets' will deliver 'a secure – adequate, affordable and reliable – supply of energy' (IEA, 2002). One practical implication of this view is that energy security does not logically imply energy independence. After all, seeking independence in a global economy would amount to protectionism and constitute an exception to the wider free trade policy. Instead of being a strategic threat, import dependency is

---

2. The economic value of lost load (VOLL) due to energy supply disruptions are critical elements of willingness to pay or demand function estimations for supply security. These estimations put a monetary value on the customers' willingness to pay for a small improvement in energy supply security at different levels of Security of Supply (SoS). Knowledge about these valuations might be of great importance for energy security policy makers when they are in need to compare the additional costs of alternative supply security investment options to the overall benefits such investments can provide for the public.

3. Energy supply security is a classic example of an externality, ie, of an issue that affects the wellbeing of individuals and society but which markets alone are not providing at adequate levels. Being a negative externality, energy supply risk constitutes a policy issue. This means private individuals cannot cover themselves for such risks due to their informational complexity and unquantifiable nature. This is where governments need to step in (NEA/OECD, 2010).

understood as just another part of the global division of labour, creating mutual benefits for exporting and importing countries alike (Keppler, 2007). Notably, this division of labour helps increase a diversity of energy sources, which is generally reckoned a stronghold in energy security (UK Government's Cabinet Office, 2002, p. 57).

8.2.3 Normative Ambiguity

Juxtaposing the economic and geopolitical perspectives reveals fundamental difficulties in attaining a single, universal notion of energy security. The two perspectives point in radically opposing directions when it comes to selecting an actual strategy: maintaining power relations versus seeking integration, seeking conflict versus seeking cooperation and allowing interdependence, framing relations as positive-sum games or alternatively as zero-sum games, etc. (Khanna, 2008). It should be clear from the above, that it is fundamentally impossible to resolve the dilemma and choose between the two. Neither of the two perspectives can be dismissed, as both highlight particular, important aspects of energy security. At the same time, neither perspective is capable of capturing the whole picture and making the other superfluous. Moreover, both perspectives neglect societal issues related to the fact that societies and the actors inside them do not only look at the aforementioned economical or geopolitical factors when considering their use of energy or, broadly speaking, their relations with the energy systems.

Nonetheless, these two perspectives complement each other by articulating different aspects of energy security challenges as well as the possible solutions (Checchi et al., 2009). On one hand, the strategic view is necessary to understand the realism of many energy policies implemented in the name of energy security, even if this view has often produced the 'mistakes and dramas' mentioned earlier in this chapter (Maugeri, 2006). On the other hand, the economic perspective is indispensable, both for the long-term explanation of energy strategies and as an essential tool for the solution of specific energy security issues.

Moreover, the relative importance of either perspective has not been the same over time and is likely to remain changing. As can be seen today, the recent growing use of geopolitical arguments in debates on energy security issues seems to reflect a renaissance of great power rivalry and a reversion to a polarised world in the early 21st century. Similarly, viewing global markets with a focus on oil imports seems now an obsolete approach: while basic economic theory would indicate that oil exporters have the capacity to cripple the economies of import-reliant countries, this hypothesis would, against current geopolitical backgrounds, require that oil producers are willing to take decisions that significantly counter their own and consumers' interests (Auerswald, 2006).

At the same time, even if the important role of geopolitics in energy markets is recognised, the role of economic laws in the long-term structural evolution of the market should not be underestimated. In a number of cases in history, the substitution of an energy source for another has been shown to be determined by economic factors, not political ones (Maugeri, 2006). This does not take away that significant examples of other determinants can be found as well, for example considerations of environmental responsibility (Wüstenhagen and Bilharz, 2006). Moreover, economic interests pushing for interdependency have been shown to be able to help prevent geopolitical tensions rather than stir violent conflict (Maugeri, 2006). While within this perspective the possibility for each country to be confident in its own security and prosperity is strictly interlinked with the strength – not the weakness – of others (Khanna, 2008), it remains to be seen how this evolves now that the relative importance of the country level of organisation, as distinct from global regions and nonstate actors, is changing.

8.3 THE IMPOSSIBILITY OF REDUCING FACTUAL UNCERTAINTY

In order to resolve the conceptual difficulties in energy security as articulated in the previous section, it is not quite helpful that no clear facts are available that could help determine which of the two perspectives brings out the most – objectively – relevant aspects of energy security. As the current section will show, empirical reality is just as multifaceted and ambiguous as is the range of conceptual takes on energy security previously discussed. Notably, reductionist strategies that attempt to capture issues of energy security into quantitative indicators are bound to miss out on aspects that are indispensable to acquiring energy security.

8.3.1 Reductionist Ontologies

As pointed out by Cherp and Jewell (2011), the current debate on measuring energy security is largely focused on finding the 'right indicator'. At large, such reductionist approaches seek to understand a complex system by reducing it to the interaction of its parts. For the sake of ontological simplicity, these parts are abstracted from the complexities of the energy system. The advantage of reductionism is that it can help operationalise a system as complex as the energy supply chain, and offer a guidance towards its assessment. In practice, reductionist thinking is all over the place for exactly these reasons. Designed to reduce complex phenomena to simple terms and functions, indicators are widely used to abstract from the energy system a few key input parameters to give an overall indication of its level of security (for an overview see Greenleaf et al., 2009; Kruyt et al., 2009).

One common strand of reductionist thinking is to explain the concept of energy security by explicitly narrowing the domain of analysis to a manageable – and, in most cases, quantifiable – level. For example, the extent of dependence of the energy system on external sources of supply could be made the primary focus. Alternatively, given a predefined set of (often implicit) assumptions about the primary determinants of energy security, reductionism can provide measurable outputs comparing the position of a country or region relative to other countries or regions. Also, such approaches could assess the contribution of new energy sources/technologies or policies.

Yet despite their ability to simplify complex phenomena, reductionist approaches suffer from some key weaknesses that challenge their actual usefulness as policy instruments. Indeed their methodologies suffer from one or more of the epistemological deficiencies (Brandon and Lombardi, 2011). Attempts at capturing the complexity of energy security using a reductionist, indicator-based approach are bound to fall short of doing justice to even part of the complexity of energy security.

In a general sense, an indicator is a signal or a proxy useful in conveying information about the state of a system. As security is commonly defined as the state of being free from threats, an energy security indicator is a signal of the state of an energy system which conveys information about its potential vulnerabilities (Cherp and Jewell, 2011). However, the complexity of energy systems hides multiple dynamic vulnerabilities, as the external boundaries of the system, its elements and their internal connections are uncertain and fluid. Therefore, it is difficult to capture all these vulnerabilities with simple indicators.

Thus, a first empirical difficulty is in scoping or defining the empirical domain of the research. Indeed, studies attempting to quantify energy security do so within different temporal, sectoral and spatial limits. In general, any empirical analysis will of course be limited, but in this case it leads to two common errors of inference. One error is that it typically leads to narrow conclusions about energy security from broad premises. The other error is that broad conclusions often obtain from relatively narrow premises.

The first error occurs when an empirically rich methodological tool – eg, an energy system model – is used to compare relatively simple energy security performance indicators (as in Böhringer and Keller, 2011; Politecnico di Torino, 2011; Criqui and Mima, 2011). Rather than attempting to characterise and explain the behaviour of the system as a whole, this approach falls into the reductionist trap of confining the output to a handful of easily measurable data points. As such, the research may account for all the interactions, technologies and processes making up a given energy system, while it only targets a subset of these – eg, the diversity of primary energy supply – for detailed analysis.

The second inferential error occurs when studies attempt to reach broad conclusions from relatively narrow premises. Despite acknowledging the complexity of the energy system, many studies often equate the security of one element in the supply chain (eg, fossil fuel imports) with the security of the energy system as a whole (Asia Pacific Energy Research Centre, 2007). Often, case studies of the energy security of a particular network or fuel do not admit their own empirical domain restriction. Indeed, a study of the impact of short-term shocks in the European gas sector (Lochner, 2011) may yield prescriptions and conclusions about energy security that are far different to a study on the long-term stresses caused by a policy-driven nuclear phase-out (Fürsch et al., 2012). A consequence of the multidimensional nature of energy security is that reduction of one type of energy security might lead to increasing the exposure to other threats (Watson, 2009).

A typical consequence of these errors is that reductionist approaches are particularly prone to viewing energy security as a distinctly supply-side phenomenon, with its root cause ultimately traced to the risk of a disruption to the smooth functioning of the primary energy supply chain. While admittedly forming an indispensable part of any analysis, this supply-side focus has been recognised as providing a partial and occasionally even misleading picture of energy security as a whole (Jansen et al., 2004). Indeed, reductionist approaches often implicitly start with the premise that a country's level of dependence on external energy suppliers constitutes the central risk to its overall energy security (Lefèvre, 2007). Often, reductionist approaches seem to uncritically equate energy security with generic values such as diversity of energy sources or the elimination of vulnerability, instead of seeing these values as part of a multidimensional conception together with energy security. Indicator-based assessments implicitly assume that diversity is the desired state for energy systems, making the degree of diversification the de facto measure of energy security. For instance, energy security has been represented by an index combining import dependency, commodity dependency and energy intensity (Bollen et al., 2010) by an indicator of diversity of primary energy sources and import dependency (McCollum et al., 2013) and by the diversity of fuel source mix in electricity, as proxy for the robustness against interruptions on any one source (Grubb et al., 2006). However, there exist many other dimensions of supply security that extend beyond the issue of diversity alone (Stirling, 2009). But the rationale for focusing on diversity is only when sources or modalities of the threats are unknown, that is a condition of ignorance. Indeed, according to Stirling (2010) energy security is related to different types of incomplete knowledge: risk, uncertainty, ambiguity, ignorance; each type requiring a different analytical armoury.

8.3.2 Disregarding Dynamics

The problem of simplifying a complex reality and reducing it to indicators is not merely in the reduction itself, if only because the argument could always be made that no single theory can capture the entirety of reality and will always need confinement. The problem is also in the fact that such reduction necessarily disregards the dynamics that emerges when a larger number of factors come into play – as is the case with issues of energy security. In particular, even if we know the probability of an incident occurring, the actual impact of the incident will be highly dependent on the constitution of the system a whole.

In the case of energy systems, the relevance of any threat is not only determined by the exposure of the energy system to the potential threat but it also depends on the system's resilience, ie, its capacity to tolerate disturbance and withstand shocks, which acts as a cushion to damper the impacts of supply-side threats (Greenleaf et al., 2009). Resilience is the ability of the system to continue to deliver affordable energy services to consumers and provide alternative means of satisfying energy service needs in the event of changed external circumstances (Chaudry et al., 2009; Gnansounou, 2008). It can be thought to be the antonym of vulnerability, which can be defined as the degree to which that system is unable to cope with selected adverse events.

As energy security is a 'property of the energy system' (Barrett et al., 2010), an adequate assessment must include all the significant elements of the system and emphasise the relations and interactions between them (Tosato, 2008). Capturing only some constituting elements of the system into indicators renders invisible the interactions and interdependencies between the technical and non-technical components of the whole energy supply chain. It is this interaction of different components, that is the integral nature of the energy system, not just its parts, that determines the system's capacity to absorb perturbations and continue energy provision (Chaudry et al., 2009; Barrett et al., 2010). Additionally, the impact of a threat does not only depend on the system as such, but may also appear differently at different timescales of observation or analysis.

Moreover, the complexity of energy systems engenders an evasive dynamics. The evolution of energy sources, technologies, markets, institutions, policies and stakeholder behaviour is constantly reshaping the structural characteristics of the system. As consequence, at different points in time the same event will impact on the system in a different manner. The cross-temporal dynamics of the system is another factor explaining its capacity to withstand adverse events (see, eg, Greenleaf et al., 2009 for a more detailed analysis of such dynamics).

By consequence, it is virtually impossible to capture the integral nature of the energy system on the basis of a partial perspective as necessarily produced by indicator-based approaches. By their very nature, indicator-based assessments of energy security are unable to capture the energy system's response to adverse events and assess the actual consequences of energy insecurity. This is because indicators cannot capture the chain of substitutions triggered by an adverse event along the whole supply chain, and hence processes such as primary energy substitutions or demand elasticities go unaccounted for. For instance, a cut in the supply of Russian gas to the EU energy system can be adsorbed through a chain of reaction along the whole supply chain, like the substitution of the source of import, the use of storage, the substitution of gas with alternative fuel in the power sector as well as in the residential sector, etc. The possibility and magnitude of any of these reactions is strictly linked to a number of factors characterising the energy system at any moment. Under certain circumstances a proper simulation of the whole chain of reactions can reveal that the actual impact of the shock could be negligible. However, as clearly stated in Greenleaf et al. (2009), even sophisticated analyses based on indicators are ultimately based on variables which are at best only proxies of the vulnerability of the system. For example, even a wide set of indicators on import dependence and the role of gas in final consumption can only provide a proxy for the vulnerability of the energy system to a physical interruption to gas imports rather than a measure of the actual disruption to imports.

Thus, it seems that indicator-based assessments, because of their methodological makeup, tend to consider the causes rather than the consequences of energy insecurity. However, this is rarely made explicit, and in many cases there is a tendency to conflate of the cause with the consequence. Indeed, many existing attempts at quantification of energy security assume, implicitly or explicitly, that the combination of variables selected to build energy security indicators is able to represent the behaviour of the energy system under an adverse event – short-term disruptions in particular. This assumption is explicitly mentioned, for instance, in Jewell (2011), Greenleaf et al. (2009) and Jensen et al. (2004). However, none of these studies actually describe the way the system would react to the potential contingencies that could affect it.

Indicator-based approaches have been supplied with additional layers of complexity. Studies attempt to systematically construct statistically robust aggregated energy security indices to account for as many variables as possible (eg, not just import diversity or price; see World Energy Council, 2014; Politecnico di Torino, 2011; Scheepers et al., 2007). Studies have also attempted to include dozens, if not hundreds, of discrete parameters categorised according to energy security's different dimensions. Sovacool and Mukherjee (2011) categorise 320 simple indicators and 52 complex indicators 'that policy-makers and scholars can use to analyze, measure, track, and compare national performance on energy security'. Martchmadol and Kumar's (2013) study presents a more statistically complex attempt at scaling and correlating 25 different indicators (eg, 'agricultural energy intensity' or 'electricity per capita') to arrive at a single aggregated index (yielding a score between 1 and 10).

Part of the rationale behind the continuous increase in the complexity of indicators is in recognition of the fact that energy security is a polysemic and multidimensional concept. An important driver is the recognition of the risk that other energy policies can affect energy security and bring about extra costs (European Commission, 2013). As a consequence, to build a secure and sustainable energy future it is necessary to assess the complex interactions between sustainability, competitiveness and security, so as to develop cost-effective strategies that maximise their results in all three policy areas.

However, notwithstanding the problems noted above, this poses a risk of conceptual conflation, which occurs when authors frontload several other dimensions of energy policy into a definition of energy security (eg, 'competitiveness' or 'sustainability'). An inconvenient paradox lies within such a practice; depending on the time horizon, it may transpire that sustainable or competitive energy may both be causes of energy insecurity. Indeed, increased sustainability may impose grid stability issues due to intermittency, while a commitment to competitively priced energy may deincentivise market players from investing in costly security measures, such as redundancy or backup capabilities. It is therefore inappropriate to include such dimensions that are in potential conflict with security without adequately recognising the synergies and trade-offs between them. Moreover, the inherent relativity tied to 'competitive' or 'sustainable' energy does not bode well for attempts at concretising the definition of energy security. Rather, such adjectives should be construed as qualifiers – secure energy that is also competitive or sustainable can be assigned a premium to secure energy that is anticompetitive or harmful to the environment.

8.4 UNSTRUCTURED PROBLEMS

After the complexities we discussed with normative and empirical perspectives on energy security, the current section will add a more substantive account of the nature of complexity itself. We will discuss different strategies that are in general available for dealing with complex issues in a policy or governance perspective. As exemplified by the two main perspectives on energy security – an economic and a geopolitics perspective – complex issues are principally ambiguous: they can be viewed in multiple perspectives, and any perspective will highlight a different subset of their constituents, thereby potentially leading to radically different assessments or conclusions (Gallie, 1956). Adding to this is what was shown in the section on indicator-based reductionism: a large part of current analyses falls short of establishing empirically the exact state of energy security, or what is needed to improve or attain it.

Following Hoppe (2002) we discern two primary axes along which uncertainty can occur: the normative and the factual dimensions (see also Ezrahi, 1980). Factual uncertainty concerns questions of how matters are, and includes ambiguities and dissensus if a sufficiently large group of relevant stakeholders de facto disagrees. If factual uncertainty is the only problem standing in the way for closure and decision, sources such as further research, expert advice, public surveys, etc. could be consulted. Normative uncertainty, conversely, regards questions such as which values are relevant and how they should be prioritised, how factual matters should be assessed against a given set of values and how values are to be realised. In case of normative uncertainty, political and moral debate could principally increase the chances to arrive at a sufficient degree of closure (though, needless to say, there will always remain a large class of normative issues on which consensus remains unattainable; just as, for that matter, not all factual uncertainty can be eliminated by further research).

Naively, it could be argued that uncertainty can be mitigated or reduced through knowledge production of some sort: if unknowns are replaced by knowns, the problem becomes principally more manageable. Normative ambiguity could be reduced by further political or moral debate on what energy security should consist of. And empirical uncertainty could be reduced by mobilising additional sources of knowledge and expertise. However, as will become clear in the following, notably the connection between normative and empirical forms of knowledge becomes problematic and renders ambiguous which kind of knowledge should be solicited in the first place.

For the sake of argument, we understand 'uncertainty' to include normative ambiguity as well as empirical dissensus. This semantic simplification is chiefly justified by the fact that both kinds of problems can be countered by increasing (some sort of) certainty. Normative ambiguity can be mitigated by acquiring moral wisdom, and empirical dissensus can be mitigated by acquiring additional scientific knowledge – or at least so in the ideal case. Also, in a more practical sense, the simplification matches the typical response to such problems: acquiring 'more certainty' is hoped to produce a further degree of 'closure' such that it can be acted upon – whether we talk about the normative or the empirical domains of uncertainty. However, as we will show here, the uncertainties merit a closer look so as to assess the strategies available for dealing with them.

Complexity emerges if both normative uncertainty and factual uncertainty occur. This is what happens in governance of transitions towards secure and low-carbon energy systems: many things with respect to causal relations, states of affairs, and possible futures are uncertain, as are the moral claims people and communities (may) make. The fundamental complexity of such problems is in the interrelation between the reduction on either axis: each of the axes of uncertainty renders attempts to improve certainty on the other axis perilous. If facts are uncertain, it is not self-evident what the political and moral debate should be conducted about, and if the moral and political programme is ambiguous, it remains questionable which further factual evidence should be sought.

These problems are variably referred to as complex, wicked, untamed or unstructured problems (Conklin, 2005; Hoppe, 2002). We prefer 'unstructured', because the approach we expose aims at first adding some 'structure' to the problem and then seeking to find a solution. Generally, seemingly straightforward strategies to reduce complexity in the case of unstructured problems merit critical attention, as they might be overpromising at best, and overly simplifying at worst. As was indeed argued in the preceding sections, both the normative and the empirical dimensions clearly resist any simplifying approach.

8.4.1 Dealing With Unstructured Problems

Generally, problems of this unstructured kind require iterative ways of knowledge production as well as reflexive forms of governance. The need for knowledge production to be iterative comes from exactly the fact that uncertainty is situated on both the normative and the factual axis: it is to some extent unclear what the problem itself is, and getting to know the problem makes up for a considerable part of finding a solution (this is one of the defining properties of wicked problems; see Conklin, 2005). Reducing both normative and factual uncertainty requires that only small steps be taken at a time, and that reduction of normative and factual uncertainty is taken equally seriously.

Solutions for complex and unstructured problems have chiefly been sought in methods to broaden the perspectives brought to bear on the problem. Disciplinary boundaries are crossed and strategies for interdisciplinary and transdisciplinary knowledge production are developed. Notably, the idea of post-normal science (Funtowicz, 2002; Funtowicz and Ravetz, 1993, 2003) has played an important part. It proposes to broaden the relevant knowledge community, and to include other perspectives than merely academic ones: it explicitly brings both factual knowledge and policy decisions under the same regime of quality and mobilises the extended peer community to warrant this quality.

In addition to an extended peer community, such strategies need iteration and learning. The relevant knowledge base is not only in the traditional academic sciences, but also in sociotechnical practice and in political reality. Consequently, knowledge production will coincide with intervention in the 'real world' as opposed to the stylised picture of knowledge production in the scientific laboratory. There is no clear-cut question that scientists or other experts can be summoned to resolve, but rather a heterogeneous composite of many questions that are to be resolved in concert. This learning-while-implementing is necessarily an iterative process, and its time dimension will prove to be an important point of connection with the conceptualisation of energy security, discussed in the following sections.

Also, solving problems of this unstructured kind requires that different sorts of knowledge and expertise are brought together. Multiple disciplines or otherwise epistemological communities[4] are needed to contribute a multitude of perspectives on the problem. As such an 'extended peer community' (Funtowicz and Ravetz, 1993) is likely to be heterogeneous and not containable in one place, learning necessarily takes place in multiple sites at the same time. Connecting those sites is not self-given but needs work instead. In addition to roping in a broader range of epistemological perspectives, the maintenance of an extended peer community also serves the purposes of creating legitimacy among a broad audience, and the purpose of enabling the address of a broader range of publics.

Learning does not only take place in terms of a body of knowledge with lacunae that are to be filled. It also relates to (at least the possibility of) inapt arrangements of institutions. The need for governance to be reflexive is in the fact that not only new strategies may have to be devised to cope with problems, but also the institutional structure of governance may need revision to facilitate those novel strategies. In order to revise such governance structures, underlying value systems may need reinterpretation, which is a profound level of learning and a very social form of learning at that (Reed et al., 2010). Also, the new institutional structure might necessarily be heterogeneous, so as to accommodate

---

4. We refer here to epistemological communities in the literal meaning of 'a community that shares a body of knowledge and a particular way of knowing that knowledge', not to the more specific term of 'epistemic communities' as international networks of experts (see, eg, Haas, 1992).

different criteria of epistemology, democratic transparency and representation, and different audiences to serve (Bijker et al., 2009; Funtowicz and Ravetz, 1993, 2001).

8.4.2 Numbers Beat No Numbers

As said in the introduction to this chapter, it turns out in practice that policy has a strong preference for propositional questions. The complex problems just discussed do not match that ideal type, and the processes of dealing with them are easily trumped by approaches that suggest that problems are more structured. This additionally explains the relative success of indicator-based approaches: even if their epistemological foundations are questionable, they produce accounts that are to a large extent 'actionable': they are sufficiently concrete to 'do' something.

Also, it is to be considered a commonplace that in policy circles, 'numbers beat no numbers'. In practice, quantified knowledge turns out to be tremendously more actionable than qualitative or interpretive knowledge. Three reasons can be identified straightforwardly. First, numbers suggest a particular unambiguity because it appears as the outcome of a clear and fixed method that is past the stage of interpretation and negotiation. Second, numbers suggest a certain hardness. Even if a confidence interval or margin of error is supplied, a number suggests that we know what we are up to, exactly because the underlying method is clear. And third, numbers suggest universality. Numbers have a remarkable capacity to survive independently of the context in which they are made (cf. the notion of immutable mobiles, Latour, 1987, p. 227). Again, the method of which they are the outcomes de facto produce a disconnection between the origin and outcome (Latour, 2010).

However, to the light of the aforementioned complexities, reducing all policy problems to mere quantitative indicators and propositional questions would produce exactly the oversimplification we pointed out in the beginning. Problems in energy transitions largely resist such simplification. Even if a problem is complex in all the senses just discussed the general governance strategies for unstructured problems can be sensibly adapted to the specific case of energy security.

8.5 GOVERNANCE FOR ENERGY SECURITY

All in all, energy security as one goal for governance resists any simple strategy along three lines. First, different normative perspectives on how the value of energy security should be specified and pursued, while each valid in their own right, produce divergent action programmes. Second, different empirical approaches, also valid each in their own right, fail to capture the whole of what matters to energy security. And third, consequently the combination of normative and empirical uncertainties makes that simple governance strategies are futile, and that reflexive and iterative governance strategies are needed. This section will offer an attempt at operationalising such a strategy for the specific case of energy security.

In order to overcome these three lines of resistance, governance of energy security could benefit from the following recommendations. First, at the institutional level, more attention should be given to the complexities underlying indicators. It would be naïve to simply call for an abandonment of simple indicators, let alone the more elaborate scaffoldings of multiindicator models. Even though they have the shortcomings articulated above, they do offer the actionable knowledge that is sometimes needed. Nonetheless, new forms of governance are to be devised that make novel connections between the realms of policy making and the realms of knowledge production. These new forms can be thought of as 'hybrid forums' (Callon et al., 2009, p. 18) in which different involved actors and different sorts of knowledge are confronted, and where it must be accepted that sometimes the best we can get is the conclusion that there are too many 'unknown unknowns'.

By consequence, at the same institutional level, it needs further recognition that indicators and models are at best partial representations, which serve as the starting point of discussions rather than as conclusions towards action. The solution of this partiality is not in adding further indicators or variables, as shown above, but in further articulation and definition of the problem at hand.

Second, at the level of normative definitions of energy security, governance is not helped by once and for all hammering out a good working definition of energy security. Rather, it should be accepted that ideas of energy security will point in multiple, diverging directions. This means that, in confrontation with empirical evidence as well as other values, it will have to be decided what perspective on energy security should be prioritised at some point. Notably, in accordance with the need for reflexive governance, decisions can only be made that have small and reversible consequences – much like what has been termed 'technologies of humility' by Jasanoff (2003). It is worth pointing out that this is paradoxical in face of the 'radical changes' that are typically called for in the context of energy transitions. While incremental changes by definition run the risk of reproducing lock-in phenomena, the alternative of a 'master plan' should be discarded because of the normative ambiguity articulated above.

Third, at the level of empirical knowledge, it should be clear that calling for further study only makes sense under specific conditions. Knowledge production should not exclusively be aimed at 'finding better indicators'. Instead, on the one hand, attention should also be directed at discussing how indicators are produced and how complexity is thereby artificially eliminated. On the other hand, it needs explicit discussion of what consequences indicators have for the normative specifications of energy security and how those are to be mobilised. Indicators may be largely unapt to directly inform policy decisions. Nonetheless, they may shed light on which elements of normative perspectives on energy security – whether economic, geopolitical or any other sort of perspective on energy security that comes with a certain degree of legitimacy – should be high-lighted or perhaps even taken out of consideration. Also, indicators may help articulate points where seemingly divergent notions of energy security are at a closer level even partially convergent.

Chapter 9 will offer an attempt at describing and operationalising possible governance strategies for energy security, on the basis of the suitable strategy to address unstructured problems described so far. The proposed approach will follow iterative ways of knowledge production to reduce both the normative and the factual uncertainty characterising energy security issues.

8.6 CONCLUSIONS

This chapter has mobilised insights from geopolitics, economics and policy studies to explore how the conundrum of governance of energy security can be approached sensibly. The answer to the problem is not in directly acquiring more knowledge, whether in an empirical sense or in a moral–political sense. Rather, it is in first taking the detour of putting the complexity of the problem out in the open, and only in second instance finding answers to small parts of the problem. The suggested piecemeal approach speaks at once to institutional changes and normative and empirical knowledge production, instead of the linear idea of 'let's first get the facts straight, and then make a decision'.

Of course, to a large extent, the complexity of energy security has been articulated before. What the current chapter does point out in an original way, though, is how different disciplines can be used in a productive way to mobilise particular perspectives on ambiguous normative issues. This insight could be further developed into situations where ambiguity is even existent within a single discipline, for example, between different schools of thought. In such cases, it might be harder to mobilise different perspectives productively, as they are more likely to compete directly for hegemony and claim to speak on behalf of the whole discipline. Also, as was mentioned at the very selection of perspectives from different disciplines, such intradisciplinary competition might unjustly feed the impression that resolution is ultimately possible. Yet, even in such situations, productive ways of mobilising the dissensus might be possible to find.

Finally, an important consequence of the whole argument of this chapter is that not only the 'input' of the problem of energy security is heterogeneous, but also the 'output' we are likely to create with it: the fact that competing normative perspectives attribute different relative importance to partial definitions of energy security not only brings out the messiness of the problem but also points towards the recognition that a good energy security strategy is most likely one that consists of various experiments that might initially appear as incompatible or working in opposing directions. In the face of the impossibility to define a problem conclusively, it might be a consoling thought that consequently there is no need for a unified solution.

REFERENCES

Asia Pacific Energy Research Centre, 2007. A Quest for Energy Security in the 21st Century. Asia Pacific Energy Research Centre, Tokyo.

Auerswald, P.E., 2006. The myth of energy insecurity. Issues in Science and Technology XXII (4).

Barrett, M., Bradshaw, M., Froggatt, A., Mitchell, C., Parag, Y., Stirling, A., Watson, J., Winzer, C., 2010. Energy Security in a Multi-polar World. Discussion Draft ESMW research cluster. Exeter University, Exeter.

Bijker, W.E., Bal, R., Hendriks, R., 2009. The Paradox of Scientific Authority: The Role of Scientific Advice in Democracies. MIT Press, Cambridge, MA.

Bohi, D.R., Toman, M.A., 1996. The Economics of Energy Security. Kluwer Academic Publishers, Dordrecht.

Böhringer, C., Keller, A., 2011. Energy Security: An Impact Assessment of the EU Climate and Energy Package. Copenhagen Consensus Center, Copenhagen.

Bollen, J., Hers, S., Van der Zwaan, B., 2010. An integrated assessment of climate change, air pollution, and energy security policy. Energy Policy 38 (8), 4021–4030.

Brandon, P., Lombardi, P., 2011. Evaluating Sustainable Development in the Built Environment, second ed. Wiley, Oxford.

Callon, M., Lascoumes, P., Barthe, Y., 2009. Acting in an Uncertain World: An Essay on Technical Democracy. The MIT Press, Cambridge, MA.

Chaudry, M., Ekins, P., Ramachandran, K., Shakoor, A., Skea, J., Strbac, G., Wang, X., Whitaker, J., 2009. Working Paper. Building a Resilient UK Energy System, vol. 23. United Kingdom Economic Research Council, London.

Cherp, A., Jewell, J., 2011. Measuring energy security: from universal indicators to contextualised frameworks. In: Sovacool, B.K. (Ed.), The Routledge Handbook of Energy Security. Routledge, Abingdon, New York.

Checchi, A., Behrens, A., Egenhofer, C., 2009. Long-Term Energy Security Risks for Europe: A Sector-Specific Approach. CEPS Working Document, 309.

Chester, L., 2010. Conceptualising energy security and making explicit its polysemic nature. Energy Policy 38, 887–895.

Conklin, J., 2005. Wicked problems and social complexity. In: Conklin, J. (Ed.), Dialogue Mapping: Building Shared Understanding of Wicked Problems. John Wiley & Sons, Chichester.

Criqui, P., Mima, S., 2011. The European Energy Sector and the Climate-Security Nexus in the SECURE Scenarios. Background Papers. Gulf Research Foundation, Geneva.

Department for Business, Enterprise and Regulatory Reform, 2007. Expected Energy Unserved. Contribution to the Energy Markets Outlook Report. Department for Business, Enterprise and Regulatory Reform, London.

European Commission, 2000. Towards a European Strategy for the Security of Energy Supply. Green Paper. COM(2000) 769 Final. European Commission, Brussels.

European Commission, 2006. A European Strategy for Sustainable, Competitive and Secure Energy. Green Paper. COM(2006) 105 Final. European Commission, Brussels.

European Commission, 2013. A 2030 Framework for Climate and Energy Policies, Green Paper. COM(2013) 169 Final. European Commission, Brussels.

Ezrahi, Y., 1980. Utopian and pragmatic rationalism: the political context of scientific advice. Minerva 18 (1), 111–131.

Funtowicz, S.O., Ravetz, J.R., 2001. Post-normal science. Science and governance under conditions of complexity. In: Decker, M., Wütscher, F. (Eds.), Interdisciplinarity in Technology Assessment, Implementation and Its Chances and Limits. Springer, Berlin, Heidelberg.

Funtowicz, S.O., Ravetz, J.R., 1993. Science for the post-normal age. Futures 25 (7), 739–755.

Funtowicz, S.O., 2002. Post-normal Science. Science and Governance Under Conditions of Complexity. Institute for the Protection and Security of the Citizen, European Commission Joint Research Centre, Ispra, Italy.

Funtowicz, S.O., Ravetz, J.R., 2003. Post-Normal science. Internet Encyclopedia of Ecological Economics. The International Society for Ecological Economics, Washington, DC.

Fürsch, L., Lindenberger, D., Malischek, R., Nagl, S., Panke, T., Trüby, J., 2012. German Nuclear Policy Reconsidered: Implications for the Electricity Market. Institute of Energy Economics Working Paper. 11/12. University of Cologne, Cologne.

Gallie, W.B., 1956. Essentially contested concepts. Proceedings of the Aristotelian Society 56, 167–198.

Giddens, A., 2009. The Politics of Climate Change. Polity Press, Cambridge (UK), Malden (MA).

Gnansounou, E., 2008. Assessing the energy vulnerability: case of industrialised countries. Energy Policy 36, 3734–3744.

Greenleaf, J.R.H., Angelini, T., Green, D., Williams, A., Rix, O., Lefevre, N., et al., 2009. Analysis of Impacts of Climate Change Policies on Energy Security. Ecofys, Utrecht.

Grubb, M., Butler, L., Twomey, P., 2006. Diversity and security in UK electricity generation: the influence of low-carbon objectives. Energy Policy 34 (18), 4050–4062.

Haas, P.M., 1992. Introduction: epistemic communities and international policy coordination. International Organization 46 (1), 1–35.

Hoppe, R., 2002. Cultures of public policy problems. Journal of Comparative Policy Analysis: Research and Practice 4, 305–326.

International Energy Agency, 2002. Energy Security. International Energy Agency, Paris.

Jansen, J.C., Arkel, W.G., Boots, M.G., 2004. Designing Indicators of Long-Term Energy Supply Security. Energy Research Centre Netherlands (ECN), Petten.

Jasanoff, S., 2003. Technologies of humility: citizen participation in governing science. Minerva 41 (3), 223–244.

Jasanoff, S., 2007. Technologies of humility. Nature 450, 33.

Kaderják, P. (Ed.), 2011. Security of Energy Supply in Central and South-East Europe. Corvinus University of Budapest, Regional Centre for Energy Policy Research, Budapest.

Keppler, J.H., 2007. Energy interdependence in a multi-polar world: towards a market-based strategy for safeguarding European energy supplies. Reflets et perspectives de la vie économique XLVI (4), 31–48.

Khanna, P., 2008. The Second World: Empires and Influence in the New Global Order. Random House, London.

Kruyt, B., Van Vuuren, D.P., De Vries, H.J.M., Groenenberg, H., 2009. Indicators for energy security. Energy Policy 37 (6), 2166–2181.

Latour, B., 1987. Science in Action: How to Follow Scientists and Engineers Through Society. Harvard University Press, Cambridge, MA.

Latour, B., 2010. On the Modern Cult of the Factish Gods. Duke University Press, Durham.

Lefèvre, N., 2007. Energy Security and Climate Policy. IEA, Paris.

Lochner, S., 2011. Identification of congestion and valuation of transport infrastructures in the Euopean natural gas market. Energy 36 (5), 2483–2491.

Lockwood, M., Kuzemko, C., Mitchell, C., Hoggett, R., 2013. Theorising Governance and Innovation in Sustainable Energy Transitions. Working Paper, 1304. Energy Policy Group, University of Exeter, Exeter.

Martchamadol, J., Kumar, S., 2013. An aggregated energy security performance indicator. Applied Energy 103 (C).

Maugeri, L., 2006. The Age of Oil: The Mythology, History, and Future of the World's Most Controversial Resource. Praeger, Westport.

Mayer, M., 2013. What Is 'geopolitics'? in Search of Conceptual Clarity. Geopolitics in the High North. http://www.geopoliticsnorth.org/index.php?option=com_content&view=article&id=45: article2&showall=1 (accessed on 13.07.15.).

McCollum, D., Krey, V., Riahi, K., Kolp, P., Grubler, A., Makowski, M., et al., 2013. Climate policies can help resolve energy security and air pollution challenges. Climate Change 119 (2), 479–494.

Miller, C.A., Iles, A., Jones, C.F., 2013. The social dimensions of energy transitions. Science as Culture 22 (2), 135–148.

National Economic Research Associates, 2002. Security in Gas and Electricity Markets. Final Report for the Department of Trade and Industry. NERA Economic Consultants, London.

Noel, P., 10 January 2008a. Is Energy Security a Political, Military or Market Problem? An Online Q & A. The Financial Times. http://www.ft.com/intl/cms/s/2/fd6ef84a-bf85-11dc-8052-0000779fd2ac.html#axzz3fmn8qCXT (accessed on 16.08.08.).

Noel, P., 10 January 2008b. Challenging the Myths of Energy Security. The Financial Times. http://www.ft.com/intl/cms/s/2/fd6ef84a-bf85-11dc-8052-0000779fd2ac.html#axzz3fmn8qCXT (accessed on 16.08.08.).

Nuclear Energy Agency/Organisation for Economic Co-operation and Development, 2010. The Security of Energy Supply and the Contribution of Nuclear Energy. NEA/OECD, Paris.

Politecnico di Torino, 2011. Risk of Energy Availability Common Corridors for Europe Supply Security. Summary Report (draft), Final Wokshop, Brussels May 13th, 2011 http://reaccess.epu.ntua.gr/LinkClick.aspx?fileticket=ez26yrScOcg%3d&tabid=721.

Reed, M.S., Evely, A.C., Cundill, G., Fazey, I., Glass, J., Laing, A., Stringer, L.C., 2010. What is social learning. Ecology and Society 15 (4), R1.

Scheepers, M., Seebregts, A., De Jong, J., Maters, H., 2007. EU Standards for Energy Security of Supply. Energy Research Centre of the Netherlands, and The Hague: Clingendael International Energy Programme, Petten.

Sovacool, B., Mukherjee, I., 2011. Conceptualizing and measuring energy security: a synthesized approach. Energy 36.

Stirling, A., 2009. The Dynamics of Security. Stability, Durability, Resilience, Robustness Energy Security in a Multipolar World. Kohn Centre, Royal Society, London.

Stirling, A., 2010. Multicriteria diversity analysis: a novel heuristic framework for appraising energy portfolios. Energy Policy 38 (4), 1622–1634.

Stirling, A., 2014. Emancipating Transformations: From Controlling 'the Transition' to Culturing Plural Radical Progress. STEPS Centre Working Papers Series. STEPS Centre, University of Sussex, Brighton.

Tosato, G., 2008. Energy security from a systems analysis point of view: introductory remarks. In: Proceedings of the 1st International Conference of the FP7 Project REACCESS. Turin.

UK Government's Cabinet Office, 2002. Energy Review Conducted by the UK Government's Cabinet Office, Performance and Innovation Unit. Cabinet Office, London.

Umbach, F., 2010. Global energy security and the implications for the EU. Energy Policy 38, 1229–1240.

United Nations Development Programme, 2004. World Energy Assessment. Overview 2004 Update. United Nations Development Programme, New York.

Watson, J., 2009. Is the move toward energy security at odds with a low-carbon society? In: Giddens, A., Latham, S., Liddle, R. (Eds.), Building a Low-Carbon Future. The Politics of Climate Change. Policy Network, London.

World Energy Council, 2014. Energy Trilemma Index. Benchmarking the Sustainability of National Energy Systems. World Energy Council, London.

Wüstenhagen, R., Bilharz, M., 2006. Green energy market development in Germany: effective public policy and emerging customer demand. Energy Policy 34 (13), 1681–1696.

Yergin, D., 2006. Ensuring energy security. Foreign Affairs 85 (2).

Chapter 9

Reducing Uncertainty Through a Systemic Risk-Management Approach

F. Gracceva[1], G. Valkenburg[2], P. Zeniewski[3]

[1]Studies and Strategies Unit, ENEA (Italian National Agency for New Technologies, Energy and Sustainable Economic Development), Rome, Italy; [2]Maastricht University, Maastricht, Netherlands; [3]Directorate-General Joint Research Centre (DG-JRC), Institute for Energy and Transport (IET), Energy Security Unit, Petten, The Netherlands

9.1 INTRODUCTION: STRUCTURING ENERGY SECURITY

As has been explained in Chapter 8, problems of energy security (ES) and climate change mitigation (CCM) are inextricably connected. They thus form complex and unstructured problems, the solution of which is hard if not impossible to accommodate by conventional structures of governance. These structures prefer to address issues as if they are clearly structured: consisting of an unambiguous problem definition, an availability of a limited set of options with clear pros and cons, and a clear identification of stakeholders. Usually, problematics are approached through indicators. On the one hand, this offers the kind of structure to problems that governance approaches need. On the other hand, indicators often offer an undue simplification.

Still, not all hope is lost when it comes to harnessing ES and CCM problematics and lining them up for governance. Even if not all ambiguities can be resolved, some reduction is still possible in the overall complexity. Our starting point is that a key issue behind the ambiguity of the transition to a low-carbon and secure energy system lies in the peculiar complexity of one corner of the problem, ie, of ES. Therefore, this chapter will develop an approach to help structuring ES problems, by addressing both the normative and empirical ambiguity described in Chapter 8. Our systemic approach to ES captures a broad variety of elements of ES in such a way that their mutual contradictions become canalised and presented less as a contradiction.

A useful first step, in order to approach sensibly the conundrum of governance of ES, is by taking the detour of putting the complexity of the problem out in the open. Indeed, ES as one goal for governance resists any simple strategy along three lines. First, different normative perspectives produce divergent action programmes.

Low-carbon Energy Security from a European Perspective. http://dx.doi.org/10.1016/B978-0-12-802970-1.00009-7
Copyright © 2016 Elsevier Ltd. All rights reserved.

For example, as seen in the previous chapter, an economic perspective is likely to instigate a strategy that secures a certain level of ES at a certain cost in a multiparty market. A geopolitical perspective, on the contrary, instigates a strategy of securing energy at the state level, leading mostly to unilateral agreements between states. Second, any single empirical approach will fail to capture the whole of what matters. Referring again to the two aforementioned perspectives, none of them is conclusively more true than the other. Either perspective has important contributions to make to political and governance decisions, and only their combination (and inclusion of yet other perspectives, for that matter) can provide adequate information. Third, because of the combination of normative and empirical uncertainties, ES problems are complex, unstructured problems. A fundamental complexity results from the interrelation between the two dimensions of uncertainty, ie, the normative and the factual one. Normative uncertainty, ie, uncertainty as to what decision should be taken, is especially precarious if there is at the same time factual uncertainty, or uncertainty with respect to what is exactly the case. And conversely, establishing what is the case is especially difficult if it is unclear what kind of decisions are at stake, since different options might call for different kinds of expertise to be enrolled (see also Chapter 8, Section 4). This entails that governance strategies based on too narrow a range of expertise are likely to be futile.

This chapter will offer an attempt at describing and operationalising possible governance strategies for ES on the basis of the suitable strategy to address unstructured problems described in Chapter 8. As suggested there, the proposed approach follows the iterative types of knowledge production to reduce both the normative and the factual uncertainty characterising ES issues: first, it takes a step to reduce the normative uncertainty, then it starts from there to reduce the degree of factual uncertainty.

The suggested piecemeal approach speaks at once to institutional changes and normative and empirical knowledge production, instead of the linear idea of 'let's first get the facts straight, and then make a decision'.

In the following section we present an argument on how a systemic approach to ES can unfold in a conceptual perspective and help reduce its normative ambiguity. In the third section we present an account of how this concept relates to empirical and factual uncertainties. In the fourth section, then, the systemic integration of ES issues are connected to the strengths and weaknesses of specific governance activities, which each have particular strengths and weaknesses vis-à-vis specific elements of systemic ES.

9.2 TOWARDS A SYSTEMIC VIEW ON ENERGY SECURITY

As discussed extensively in Chapter 8, ES is normatively problematic. However, even if it is difficult to reach a generally acceptable and accepted definition of ES (Sovacool and Mukherjee, 2011) a review of the literature makes it possible to identify some common traits that many definitions and/or conceptualisations seem to share (see Table 9.1, below, which is a revised and extended version of the review of definitions collected by Winzer, 2012).

TABLE 9.1 Definitions of Energy Security – A Few Recurring Key Elements

Author (Year)	Risk	Multi-dimensionality	Relativeness	System Response
O'Leary et al. (2007)	✓	Geographical distribution of primary fuels and to the operational reliability	Tolerable level of risk, affordability	
Bertel (2005)		Volume Price	Lack of vulnerability	Lack of vulnerability
Checchi et al. (2009)	✓	Risks that derive from energy use, production and imports	Cost-effectiveness	

O'Leary et al. (2007): 'a country's energy security policy generally comprises measures taken to reduce the risks of supply disruptions below a certain tolerable level. Such measures should be balanced to ensure that a supply of affordable energy is available to meet demand. Security of energy supply thus encompasses both issues of quantity and price. However, time is also a key parameter, (...). Insecurity in energy supply originates in the risks related to the scarcity and uneven geographical distribution of primary fuels and to the operational reliability of energy systems that ensure services are delivered.'

Bertel (2005): 'The notion of security of energy supply... may be defined in a broad sense as the lack of vulnerability of national economies to volatility in volume and price of imported energy.'

Checchi et al. (2009): 'security of supply is essentially a strategy to reduce or hedge risks that derive from energy use, production and imports. All security-of-supply approaches are aimed at "insuring" against supply risks with an emphasis on cost-effectiveness and the shared responsibility of governments, firms and consumers.'

Continued

TABLE 9.1 Definitions of Energy Security – A Few Recurring Key Elements—cont'd

Author (Year)		Risk	Multi-dimensionality	Relativeness	System Response
Creti and Fabra (2007)	'In the short-term, supply security requires the readiness of existing capacity to meet the actual load; supply adequacy; instead, refers to the long-run performance attributes of the system in attracting investment in generation, transmission, distribution, metering, and control capacity so as to minimize the costs of power supplies.'	✓	Security Adequacy Generation Transmission Distribution	Readiness/adequacy to minimise costs	
Department of Trade and Industry (2002)	'No energy form and no source of supply can offer absolute security, so improving security of supply means reducing the likelihood of sudden shortages and having contingency arrangements in place to limit the impact of any which do occur.'	Reducing the likelihood of..., contingency arrangements		Limit the impact	✓
Doorman et al. (2006)	'System vulnerability, which is defined as the system's inadequate ability to withstand and unwanted situation.'	Vulnerability		Adequacy	System's ability to withstand
Grubb et al. (2006)	'Security of supply, for the purposes of this paper it can be defined as a system's ability to provide a flow of energy to meet demand in an economy in a manner and price that does not disrupt the course of the economy. Symptoms of a non-secure system can include sharp energy price rises, reduction in quality (eg, brown-outs), sudden supply interruptions and long-term disruptions of supply.'		Price rises, reduction in quality, supply interruptions and long-term disruptions	Manner and price that does not disrupt the course of the economy	System's ability to provide a flow of energy

IEA	'Energy security is defined as the availability of a regular supply of energy at an affordable price (IEA, 2001).' Security of supply consisting of adequacy, reliability and affordability.		Adequacy, reliability, affordability	Regular, affordable	Ability of an economy to guarantee
Intharak et al. (2007)	'Energy security as the *ability of an economy* to guarantee the availability of energy resource supply in a sustainable and timely manner with the energy price being at a level that will not adversely affect the economic performance of the economy. (...) there are 3 fundamental elements of energy security (...): (1) physical energy security; (2) economic energy security; and (3) environmental sustainability.'		Physical, economic, environmental	Level that will not adversely affect the economic performance	
Joskow (2005)	'it is useful to group "supply security" concerns into two categories: (a) short run system operating reliability and (b) long run resource adequacy.'	✓		Reliability/adequacy	
Kaderják (2011)	'Policy for supply security must primarily strike a balance between the costs of improving supply security and the benefits from it in the form of a reduction of welfare loss owing to non-provided energy services.'	✓		✓	

Continued

TABLE 9.1 Definitions of Energy Security – A Few Recurring Key Elements—cont'd

Author (Year)		Risk	Multi-dimensionality	Relativeness	System Response
Keppler (2007a)	*'Traditional* definitions of energy supply security combine a short-term notion of the continuity of physical supplies with long-terms notion of "affordable" prices, "competitive prices" or "adequate prices". The risk management approach to the security of energy supplies argues that supply security is an issue dependent on the risk-adverseness of consumers. Its focus is thus not the absolute level of energy prices but the size and impact of changes in energy prices.'	Risk-adverseness of consumers	Continuity versus adequacy Affordable versus adequacy	Focus is thus not the absolute level	Focus on size and impact
Kruyt et al. (2009)	'For energy security, ...elements relating to SOS: availability – or elements relating to geological existence. Accessibility – or geopolitical elements. Affordability – or economical elements. Acceptability – or environmental and societal elements.'		Availability Accessibility Affordability Acceptability		
Lesbirel (2004)	'Energy security, like the concept of security itself is a contestable concept. Rather than seeking to define energy security comprehensively and while acknowledging different conceptions of it, I stress the notion of insurance against risks.'	Insurance against risks			

Continued

Lieb-Doczy et al. (2003)	'Security of supply is fundamentally about risk. More secure systems are those with lower risks of system interruption.... Our adopted general definition of energy security (no major frictions to the economy caused by the energy system)...'	SoS is about risk		No major frictions	√
McCarthy et al. (2007)	'Security includes the *dynamic response of the system* to unexpected interruptions, and its ability to endure them. Adequacy refers to the *ability of the system to supply* customer requirements under normal operating conditions.'		Security Adequacy	Adequacy	*Dynamic response of the system, ability to endure*
NEA/OECD (2010)	'Security of energy supply is the resilience of the energy system to unique and unforeseeable events that threaten the physical integrity of energy flows or that lead to discontinuous energy price rises, independent of economic fundamentals.'		Energy flows, Energy prices	Integrity, Continuity	Resilience
NERA (2002)	'Energy security of supply is generally expressed in terms of risk. An energy system is secure enough when the risk that consumers will be unable to access energy supplies at prices that reflect the cost of provision is sufficiently low. (...) Security always need to be seen in terms of system impact. (...) All electricity and gas systems provide *some* level of security to consumers. (...) whether any given level of security is optimal or even adequate is not easy to assess. The idea of risk is a good starting point. There will be a trade-off between security and costs.'	√	Short-term operation of the system, long-term incentives to investments	An energy system is secure enough when risk is sufficiently low	Security always need to be seen in terms of system impact

TABLE 9.1 Definitions of Energy Security – A Few Recurring Key Elements—cont'd

Author (Year)		Risk	Multi-dimensionality	Relativeness	System Response
Rutherford et al. (2007)	'The term energy security to refer to a generally low business risk related to energy with ready access to a stable supply of electricity/energy at a predicable price without threat of disruption from major price spikes, brown-outs or externally imposed limits.'	Low business risk related to energy		Stability Predictability	
Stern (2002)	'There are two major dimensions of these risks: short-term supply availability versus long-term adequacy of supply and the infrastructure for delivering this supply to markets; operational security of gas markets, ie, daily and seasonal stresses and strains of extreme weather and other operational problems versus strategic security, ie, catastrophic failure of major supply sources and facilities.'		Availability Adequacy Security Strategic		
Wright (2005)	'Security of gas supply: an insurance against the risk of an interruption of external supplies.'	Insurance against risk		Adequacy	

First, ES is generally presented as concerned with 'risks'; operationally, this implies that the key point is to determine the degree to which different risks should be mitigated, an arduous task when many risks are difficult to quantify and in some cases even to identify (NERA, 2002; Keppler, 2007a; Chester, 2010; Winzer, 2012). Second, ES is a multifaceted concept as the supply chains making up the energy system are subject to a wide range of different threats, and vulnerabilities along the energy supply chain are closely connected to spatial and temporal contexts, and specific market situations (Chester, 2010).[1] These risks and their acceptability are often broken down into their exposure on the availability of energy and the price at which energy is available (EC, 2000; Grubb, 2006; Checchi et al., 2009).[2] Third, ES is an intrinsically relative concept, as the degree of security of an energy system can be defined only through relative concepts, eg, adequate physical availability of energy or affordable prices.[3] Last, the actual level of security of an energy system depends on the actual capacity of the potential threats mentioned above to negatively affect the normal functioning of the energy system and to determine major frictions to the economy caused by the energy system and ultimately produce a loss of welfare (Bohi and Toman, 1996; DTI, 2002).

In this context ES is the product of the interactions and interdependencies of a complex system, ie, one whose properties are not fully explained by an understanding of its component parts: ES becomes a property of complex and dynamic energy systems (Barrett et al., 2010; Tosato, 2008; Helm, 2003) and it 'always needs to be seen in terms of system impacts' (NERA, 2002, p. 6).

Based on the points above we can define a secure energy system as one that is evolving over time with the adequate capacity to satisfy the energy service needs of its users under any circumstances, ie, even when affected by unforeseeable events that threaten the physical integrity of energy flows or that lead to discontinuous prices of energy services (Gracceva and Zeniewski, 2014; Keppler, 2007a,b). This definition implicitly includes the traditional view of ES as 'the uninterrupted availability of energy sources at an affordable price'.[4] At the same

---

1. As such, ES is in fact an all-encompassing phrase for a number of issues as diverse as operation of energy networks, availability of alternative sources of energy, cost-oriented prices and market power in national and international energy markets, and investments in a wide array of energy infrastructures.
2. The main risks for the energy system are the interruption in the physical availability of energy products (EC, 2000) and energy prices that can disrupt the course of the economy (Grubb, 2006), because if prices were allowed to rise without limit, there would always be a sufficiently high price at which demand would equate supply (Checchi et al., 2009); the relative importance of the two quantity and price components can vary depending for instance on the market structure.
3. For instance, according to Mitchell (2009) these concepts leave open questions like, 'How much physical availability – anything consumers want, now and in the future?' or, 'What is an affordable price? Affordable by whom?'
4. See, eg, http://www.iea.org/topics/energysecurity/. The same definition is adopted in the EU Green paper (COM (2000) 769 final): the EU's 'long-term strategy for energy supply security must be geared to ensuring, for the well-being of its citizens and the proper functioning of the economy, the uninterrupted physical availability of energy products on the market, at a price which is affordable for all consumers (private and industrial)'. However, the EU definition is expanded to 'environmental concerns and sustainable development, as enshrined in Articles 2 and 6 of the Treaty on European Union'.

time this definition captures the multidimensionality and complexity of ES and is helpful for reducing the normative ambiguity of the concept. Indeed, by identifying a few key elements characterising the problem of ES, this multidimensional and systemic take helps to identify the values which need to be assessed, how they should be prioritised and how factual matters should be assessed against a set of values. This helps to give a first structure to what we have defined above as an 'unstructured problem' and through that it also helps to make progress on the other axis along which uncertainty occurs, ie, the factual uncertainty. The next section will take advantage of the progress made on the normative dimension to propose a framework addressing the dimension of factual uncertainty.

9.3 SECURE ENERGY SYSTEMS IN PRACTICE

Having questioned the conceptual and methodological utility of 'reductionist' methodologies (Chapter 8), we propose an alternative paradigm which reflects the 'polysemic' and multidimensional nature of ES and helps making it a 'partially structured' problem.

Following Stern (2004), this alternative paradigm aims to be a systemic framework that can be used 'to analyze the impact of specific security events, the level of risk attached to such events, and the cost of measures which would provide insurance against them'. Upon the systemic perspective to ES provided above it becomes clear that 'energy security is no longer about the uninterrupted provision of a particular form of energy. Rather, the key policy task becomes to design a framework for insurance and for allocating risk efficiently between private players and public players, knowing that markets cover *risk* very well and *uncertainty* very badly' (Keppler, 2007a, p. 22).[5] As a consequence, the wide range of factors that can exercise a stabilising influence on the energy services delivery system must be identified and thoroughly analysed, together with their relations and interactions/synergies. In the following section, we present a strategy for doing this, resulting from a review of the wide literature on ES issues,

---

5. Following Stirling (1999, pp. 16–17) 'the well-established formal definition of *risk* is that it is a condition under which it is possible both to define a comprehensive set of all possible outcomes and to resolve a discrete set of probabilities (or a density function) across this array of outcomes. This is the domain under which the various probabilistic techniques of risk assessment are applicable, permitting (in theory) the full characterisation and ordering of the different options under appraisal. The strict sense of the term *uncertainty*, by contrast, applies to a condition under which there is confidence in the completeness of the defined set of outcomes, but where there is acknowledged to exist no valid theoretical or empirical basis for the assigning of probabilities to these outcomes. Here, the analytical armoury is less well developed, with the various sorts of scenario analysis being the best that can usually be managed'. Sterling then adds a further category, ie, *ignorance*, which 'applies in circumstances where there not only exists no basis for the assigning of probabilities (as under uncertainty), but where the definition of a complete set of outcomes is also problematic. In short, it is an acknowledgement of the possibility of surprises. Here, it is not only impossible to rank the options but even their full characterisation is difficult. Under a state of ignorance (in this strict sense), it is always possible that there are effects (outcomes) which have been entirely excluded from consideration'.

characterised by a common feature: the reviewed studies are mainly quantitative and share the ambition to provide practical insights on specific ES strategies based on rigorous assessments.

The main point to draw from this literature review is that it is very important to have a methodology that clearly identifies the system under consideration, the specific risks affecting this system, and the criteria for assessing them (eg, measure of damage, likelihood/consequences). Indeed, the three steps identified by Stern (2004) have a clear similarity with the traditional structure of a risk assessment approach, which is also a coherent with the framework proposed by the European Commission (EC) communication (see Table 9.2) to deliver the internal electricity market and make the most of public intervention (C(2013) 7243 final). This document is important for the current discussion because it addresses the urgent need expressed in the Green Paper on a 2030 framework for energy and climate policies to review public intervention in order to move towards a more sustainable, secure and competitive energy system for the longer term.

Similarly, the EC regulation on security of gas supply (EC/994/2010) has introduced the need for any Member State to produce a periodical risk assessment on its security of gas supply, following the structure of traditional risk assessment, as suggested by the Joint Research Centre Institute for Energy and Transport (Bolado-Lavin et al., 2012).

In the following paragraphs we briefly describe and discuss the key steps of a risk management approach to ES, by drawing from the literature one or more examples for each step.

9.3.1 Domain Specification

9.3.1.1 Establishing the Context and Risk Identification

The preliminary step of any risk assessment is to establish the context, which requires an articulation of the central research questions as well as the scope and characteristics of the system under study.

As said above, energy systems are subject to a wide range of different threats and vulnerabilities along the energy supply change are closely connected to spatial and temporal contexts: on one hand, threats may arise in the mining and extraction of energy, through its production, transport and conversion, all the way down to its final use; on the other hand, transient disruptions (eg, shocks like a sudden interruption of some gas supplies to the European Union) can be differentiated from more enduring pressures (eg, stresses like a prolonged limited access for foreign investors to the resources of energy rich countries), which compromise the long-term ability to deliver energy supplies to end-users at cost-oriented prices. Therefore, the preliminary problem of ES assessments is the choice of the temporal, spatial and substantive boundaries of the system under analysis (Cherp and Jewell, 2011). Within each level of analysis, different forms of risk appear. This means that enlarging or minimising the level of

TABLE 9.2 Reflection of a Risk Management Approach to Energy Security on Key European Community Documents on Gas and Electricity Markets

	Risk Assessment Approach (ISO 31000)	C(2013) 7243 Final, Section III: Making Public Intervention More Effective and Efficient	European Community (EC) Regulation Concerning Measures to Safeguard Security of Gas Supply (EC/994/2010)
Risk assessment	• Establishing the context: articulation of research questions as well as scope and characteristics of the system under study	• Identifying a specific problem and its cause	• Description of the energy system under study, in terms of its market characteristics, infrastructures and infrastructure utilisation, institutional context • Geographical scope of the system and level of detail of its description
	• Risk identification: identify sources of risk, areas of impacts, events and their causes		• Identification and description of the potential events threatening the system
	• Risk analysis: consideration of the causes and sources of risk, their positive and negative consequences, and the likelihood that those consequences can occur	• Assessing potential interplay with other policy objectives: MSs should plan holistically taking into account all objectives of energy policy (...) Trade-offs may be complex	• Definition of scenarios, including probability estimation if possible, and assessment of consequences at national and regional level, in terms of physical provision of natural gas, economic impact, environmental impact of potential security measures
	• Risk evaluation: to assist in making decisions, based on the outcomes of risk analysis, about which risks need treatment and the priority for treatment implementation	• Demonstrating that the internal electricity market functioning on the basis of the existing acquis of Union law is unlikely to solve the problem • Evaluating alternative options: the Union law acquis may offer alternatives to public intervention. Minimising impacts of public intervention	• Construction of a risk matrix

- Risk treatment: to select one or more options for modifying risks and implementing those options

- Keeping costs low: (...) MSs should ensure that the intervention is appropriate to the objective pursued and does not go beyond what is necessary
- Considering the impact on costs for consumers
- Monitoring, evaluation and phasing out of support

analysis means changing one's risk criteria. For example, if the context is the energy system, then the risk criteria encompass any event which affects the system's total capacity to deliver energy services to end-users (Chaudry et al., 2009). If the perception of risk is extended such that it signifies a loss of welfare as a result of a change in the price, supply or demand for energy services, then the context becomes the macroeconomic system and its flexibility in substituting energy services with other factors of production.

Choosing the domain of analysis within which to explore ES is simply the first step in a structured process of assessing and managing risk. The second step – that of risk identification – prompts the researcher to consider the question, 'What can go wrong and why?' These first steps together address the risks of domain under-specification and fallacy of equivocation (see above, Chapter 8, Section 3). On the one hand, quantitative assessments on ES often incur in two common errors of inference, that is, to reach narrow conclusions about ES from broad premises, the other making broad conclusions from relatively narrow premises. On the other hand, many studies implicitly assume ES to be synonymous with, rather than conditioned by, other terms such as diversity or vulnerability, while caution must be exercised when arguing for conceptual equivalence of such terms when their meanings, however subtle, are qualitatively different.

The number of threats that are caused by or have an impact on the energy supply chain is so wide (Gnansounou, 2008) that often the main reason for difference between ES concepts is the way in which the authors select the subset of these threats that they consider in their analysis. Studies on ES focus on different risk sources. On one hand, there are a few sophisticated quantitative assessments of degree of security of specific energy markets which start from a clear identification of the potential threats for those markets, eg, OXERA (2007) and POYRY (2010) for the UK gas market, de Joode (2006) for the Dutch oil sector. On the other hand, there are some nonquantitative studies which attempt to identify and categorise the main threats to the security of the whole energy system, structuring them in multiple ways – eg, by origin, intensity, dimension and so on (see Flouri et al., 2008; Huntington, 2005; IEA, 2009 (Intharak et al., 2007); Checchi et al., 2009). Adapting a framework developed by Stirling (2009a), Gracceva and Zeniewski (2015) have proposed a categorisation of potential threats to the security of the energy system on the basis of three primary dimensions (these three elements are also important for assigning probabilities and considering consequences of identified risks): location in the energy supply chain, temporality and provenance.[6]

---

6. Whereas location and temporality are the two attributes defined above, while provenance differentiates between events originating inside or outside the energy system under study, something that will affect the degree of control over their origin: internal events can be 'controlled' in the sense that there is a freedom to choose strategies that affect the likelihood of the threats; on the contrary, in case of external events the main strategy available is to 'respond' in ways that maintain the level and quality of energy services or improve the capacity of the system to adapt to events (an example is the increase of short-term flexibility through storage).

Taken together, these dimensions are designed to enable comprehensive categories with which to identify risks with as much analytical precision as is necessary to perform the proceeding steps of the risk assessment. Moreover, the identification of the wide range of possible one-event impulses and gradually strengthening (or declining) forces exercising a negative (or positive) impact on the energy system helps clarify the multifaceted nature of ES. This is important because it makes it possible to put any threats into a wider context and because the experience shows how energy crises often have more correlated root causes (see for instance Weare, 2003 on the California electricity crisis). As a consequence, also the strategies to mitigate one type of ES threat can heighten exposure to other threats (Watson, 2009).

9.3.1.2 Risk Analysis: Likelihoods and Consequences of Energy Security Risks

The third step of our paradigm involves the analysis of the previously identified risks in terms of their likelihoods and consequences, in order to analyse the actual relevance of these risks, by considering the nature of their probabilities and consequences, and in so doing to identify the attributes/properties of the system that enable it to cope with the identified threats. This will address the risk of focusing on proxies, ie, on the causes of energy insecurity instead of its consequences (see again Chapter 8, Section 3).

9.3.1.3 Likelihoods of Energy Security Risks

In order to adequately assess the likelihood and consequences of threats, the risk assessment framework clearly sets the need to reflect on the different nature and wide variety of events which can threaten the energy system. Stirling (1999) provides a useful way to classify different types of incomplete knowledge ('incertitudes', according to his lexicon). He speaks of risk if the nature of the shock or stress can be defined and a quantitative probability can be attached to it. An example in our context would be the acute shutdown of a major power plant. Percentages of 'uptime' are largely specified for such plants, and the consequences of their 'downtime' can largely be charted. Uncertainty is the case if the nature of the shock or stress can be defined but no probability can be attached to it (eg, a breakdown of global oil markets that disrupts supply). Its consequences can be roughly estimated, but we do not know when or at which probability it will happen. It is called ambiguity if there is a basis for probabilities but outcomes are not well defined. This has been the case with gas drillings in the Netherlands: there has always been some likelihood that soil would be destabilised, what the consequences would be was uncertain for a long time. And finally, ignorance exists if neither the nature of the shock or stress can be defined nor a quantitative probability can be attached to it. An example is the terrorist attack to the Twin Towers (September 2001) and its consequences on the global equilibrium.

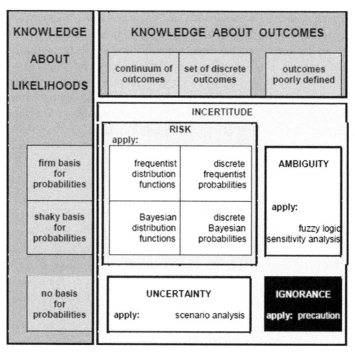

FIGURE 9.1 The concepts of 'incertitude', 'risk', 'uncertainty' and 'ignorance' (after Stirling). *Reproduced from: Stirling, A., 1999. On Science and Precaution in the Management of Technological Risk, Volume I – A Synthesis Report of Case Studies, an ESTO Project Report Prepared for the European Commission. JRC Institute Prospective Technological Studies, Seville, p. 17.*

According to Stirling, each type of 'incertitude' requires a specific analytical armoury (Fig. 9.1). Stirling's categorisation is helpful in selecting the appropriate methodology to be used to analyse different threats. For example, the use of probabilistic risk assessment techniques is only appropriate when there is comprehensive knowledge of all possible outcomes with an accompanying discrete set of probabilities. This level of certainty is typically only available for the threats characterised by statistical variability, for which data can be collected and used. Enlarging the scope of analysis beyond this dimension precludes the use of purely quantitative risk assessment. Indeed, when there is knowledge of outcomes but not of probabilities (ie, 'uncertainty', as in a geopolitically motivated cessation of energy supplies), scenario analyses can be considered a more appropriate tool.

9.3.1.4 Consequences of Energy Security Risks

The second fundamental component of the condition of risk is the actual impact produced by a threat, once it materialises. Whereas the probability of threats may vary independently of the constitution of the system, their actual impact is contingent on the structural characteristics of the system present during the timeframe in which the threat is manifest, ie, on its ability to withstand the effects thanks to its capacity to cope with the event. A systemic approach is built around the capacity of the energy system to tolerate disturbance rather than focusing on single measures, since the security of an energy system 'shall be judged by its ability to withstand shocks and to adapt', which implies that 'the resilience (flexibility, elasticity) of the system thus becomes key'.

Assessing the capacity of an energy system to tolerate disturbances needs a simulation of the behaviour of the system in response to contingencies. As consequence, the wide range of factors that can exercise a stabilising influence on the energy services[7] delivery system must be identified and thoroughly analysed, together with their relations and interactions/synergies (Jansen, 2009).[8] A system-level approach seeks to take into account as much of this complexity as possible.

Moreover, as threats to ES can come in many forms, a 'secure' energy system must possess multiple behavioural traits, ie, a number of different types of properties that act as coping mechanisms to manage such threats. Considering the location, temporality and provenance of various threats we identify five separate traits (Table 9.3), namely, stability, flexibility, resilience, adequacy and robustness, each of them associated to the categories of threats identified in Gracceva and Zeniewski (2015). These systemic coping mechanisms are centred around the actual capacity of any adverse event to affect the energy system, which depends on the way the latter can cope with this event, and is contingent on the structural (physical and nonphysical) characteristics of the system at the time it is affected by the event, ie, on the interaction of different technical and economic interconnected components working along its whole value chain. Clearly, conceding the systemic approach to energy described in this section effectively means eschewing explanations based on a handful of assumptions about the level of supply diversity or the political risks of energy dependency.

9.3.1.5 Risk Evaluation: Strategies to Enhance Energy Security

The fourth step of a risk management approach to ES is to determine if the degree of security currently provided by the system is adequate or must be corrected; in the latter case alternative strategies to increase the security of the system must be compared. This step addresses the need of a clear criterion to select ES strategies and to answer practical concerns which are directly relevant to policymakers.

ES assessments should provide insights on the 'optimal' (or at least appropriate) level of security, whereby the marginal benefit of its further increase would equal its marginal cost. In theory, the way to harmonise the metrics of all the target variables is by using monetary values. Estimations on the monetary value of the customers' willingness to pay for a small improvement in ES might be of great practical utility for policymakers, as they would enable a comparison

---

7. Energy service needs are those provided as useful energy, ie, the light, heating, cooling, power and mobility services that are used at the very end of the supply chain.
8. Whereas the energy system can be as wide as to include 'the set of technologies, physical infrastructure, institutions, policies and practices located in a defined territory which enables energy services to be delivered to its end-users' (Chaudry et al., 2009).

TABLE 9.3 A Summary of the Main Energy Security Dimensions

	Definition	Threats
Stability	Capacity of the highly interconnected energy system to maintain its operation within acceptable technical constraints.	Sudden disruptions of critical system components
Flexibility	Ability of a system to cope with the short-term uncertainty of energy system variables, by balancing any deviations between the planned or forecast supply and demand, on one side, and the actual outturn in real time, on the other side.	Statistical variability of energy variables
Resilience	Ability of the energy system to source alternative modes of production or consumption in response to sudden and transient shocks, eg, the interruption of a major supply source.	Sudden/transient disruption, deliberate use of market power
Adequacy	Reasonable expectation that the system as a whole is able to meet all demand at all times under all anticipated conditions, taking into account market conditions and the regulatory regime.	Market failures, faulty market design
Robustness	Actors in energy market are allowed to choose from primary energy sources at cost-oriented prices, without being hindered in their choice by economic or (geo) political constraints on energy resources/ infrastructures.	Enduring pressure on energy resources/ infrastructures hindering choice of energy

with the additional costs of alternative investment options (Kaderják, 2011). This is especially relevant since the financing of such often very expensive investments (eg, new transmission lines, upgrades of underground natural gas storage sites) mostly involve some form of public payment (Kaderják, 2011). The calculation of the consumer's willingness to pay for extra security[9] could make it possible even to select the 'optimal' level of security of the system. Unfortunately, this function requires knowledge of both the demand and the supply curves of extra security, or a calculation of the marginal damage function of supply disruptions (NERA, 2002; Damigos et al., 2009; Hirschhausen et al., 2008; Kaderják, 2011).

---

9. The definition and measurement of the economic value of lost load (VoLL) due to energy supply disruptions are critical elements of willingness to pay or demand function estimations for supply security. These estimations put a monetary value on the customers' willingness to pay for a small improve in energy supply security at different levels of security of service (SoS) (Kaderják, 2011).

However, a wide number of studies have been based on the estimation of the *value of lost load* (VoLL; eg, DBERR, 2007a,b and de Nooij et al., 2007) to assess the optimal level of security in electricity markets. Indeed, several electricity markets around the world establish reliability standards based on estimations of VoLL. An example is the Australian national electricity market (NEM), which is required to operate within defined levels of reserves in order to meet the required standard of supply reliability. When reserves acquired by the Australian market operator are dispatched they are bid in at VoLL, thus setting the spot price at the maximum level. Moreover, cost-benefit analyses are a typical key component of the methodology followed by transmission system operators (TSOs) to decide on investments on transmission lines, which are strongly linked to security of supply considerations (see, eg, de Nooij, 2011 and ENTSO-E documents).

Similarly, Kaderják (2011) develops an estimation of the value of gas supply interruptions for gas customers in Hungary by drawing from the experiences that industrial gas customers gained during the grand 'experiment' of the 2009 January gas crisis. The study illustrates the usefulness of VoLL estimates by providing a cost-benefit analysis of a much debated strategic gas storage investment that was commissioned in 2009 in Hungary. Interestingly, the result of the analysis is that this investment, backed from policymakers, was not justified from an economic point of view.

Oxera (2007) and Poyry (2010) are two interesting examples of estimations of the economic impact of energy insecurity. In both studies the estimated expected outages produced by some potential adverse events are used to quantify the monetary losses arising from by these events. The final result is a cost-benefit analysis of a set of potential security of supply measures, based on their contribution in terms of the Net Present Value of reduced cost of expected demand loss.

When the estimated cost of energy insecurity is calculated, it is also possible to assess the cost-effectiveness of remedial measures, by comparing the costs and benefits of alternative strategies to increase ES. There are several examples of studies selecting the economically rational strategies to increase ES. For instance, de Joode et al. (2006) assess the costs and benefits of a set of possible security of supply measures against a baseline 'disturbance-free' scenario. The benefits are measured against the costs incurred by a disruption.

Finally, a wide literature has also focused on the cost of oil insecurity, through estimations of the military cost of securing energy, with a particular focus on the US military expenditures to protect the use of Persian Gulf oil (see Delucchi and Murphy, 2008; Stern, 2010; Dancs, 2008).

9.3.1.6 Risk Treatment and Coherence With Other Energy Policy Objectives

Once risk analysis and evaluation leads to the conclusion that the security of the system should be improved, the final step is to put in place appropriate strategies. Here, the distinction between internal and external threats (Stirling, 2010) has direct implications on the types of strategies that can be put in place in the different cases. Internal events can be 'controlled' in the sense that there is a freedom to choose strategies that affect the likelihood of the threats, eg, by reducing energy demand growth, changing the market design or enlarging its size (which reduces the variability of energy system variables). In case of external events, on the other hand, the main strategy available is to 'respond' in ways that maintain the level and quality of energy services or improve the capacity of the system to adapt to events; examples are the increase of short-term flexibility through storage and the diversification of energy sources or more generally the attempt to change the long-term trajectory of the system (eg, by lowering exposure to dependence on outside sources for energy supplies or technologies).

Risk treatment strategies must also take into account other energy policy goals (Watson, 2009). After all, the literature often identifies three necessary pillars of any comprehensive energy policy – namely, security, sustainability, and competitiveness (eg, COM (2013b) 169 COM (2010) 639, COM (2006) 105). However, there are only a few analyses which have been built explicitly to integrate the assessment of ES with the assessment of these other energy policy objectives, most of which are indicator-based analysis of the interaction between climate and ES objectives (Greenleaf et al., 2009, Lefèvre, 2007). As such, these studies have all the methodological shortcomings discussed in Chapter 8 (Section 3).

The choice of risk treatment strategies raises the question of whether markets succeed or fail in efficiently producing an adequate level of secure supply of energy. Indeed, as a general economic principle governments should intervene with security of supply only if energy markets fail to realise efficient solutions (Bohi and Toman, 1996), ie, only if such an intervention can improve social welfare (de Joode et al., 2006). If there would exist perfectly competitive and informed markets for security this optimum would be reached as a result of individual decisions. On the contrary, where there are market failures, government intervention may seek to rectify these through regulation, economic instruments or other forms of intervention. After all, security of supply can be viewed as a problem of externalities, ie, costs or benefits that are ignored by markets in the determination of prices. Moreover, energy system resilience is not a pure private good, as it is 'a system property with important public goods characteristics' (Helm, 2003, p. 260). In light of these complications the treatment of ES risks are best derived from a 'framework for insurance and for allocating risk efficiently between private players (quantifiable risk) and public players (non-quantifiable risk or uncertainty)…given that markets cover risk very well and uncertainty very badly' (Keppler, 2007a,b).

9.4 GOVERNANCE OF SYSTEMIC RISKS

In the sections in this chapter, an operation on ES has been performed that offers further structure to the problem of ES without unduly simplifying it. The last step we perform in the current section consists of using the imposed structure to map issues of ES onto various activities of governance that each serve particular sociopolitical goals.

The transformation offered by the systemic approach to ES and its ensuing focus on risk is of a particular kind. It does not unduly reduce the complexity of the problem, as it leaves in place all the elements that constitute ES. Rather, it offers a structure that can be thought of as 'orthogonal' to these constituting elements: the way risks are spread over society. The way they are arranged into various temporal and spatial categories, leaves in place the many aspects by which the constituting elements can be meaningful.

A corollary of this systemic approach is that, as energy systems are dynamic, the security of a system is also a dynamic concept: at any point in time the same event will impact on the system in different measure. Also, ES is essentially about how vulnerabilities will develop in the future, not the present or the past (Cherp and Jewell, 2011). This means that iteration and learning are indispensable parts of governance of ES. The above approach to the problem helps specify at which temporal scale this iteration needs to take place. It thus helps set agendas for particular sites of learning and for specific realms of governance.

9.4.1 Different Modes Have Different Temporalities

The division of ES into various temporal scales (viz. stability, flexibility, resilience, adequacy and robustness) corresponds to a similar, be it rough, division of what is both required and possible in governance. While it has been argued (Brown, 2009) that democratic governance is not a singular arrangement but rather something that requires an ecology of institutions that each have their own strengths and weaknesses vis-à-vis particular problems, fairly little has been done as yet when it comes to relating specific institutions to specific temporalities. Yet, after the elaboration above, it is fairly straightforward that the need for democratic accommodation is different for issues that must be resolved within minutes than it is for things that maybe will take generations to implement and that currently have no clearly defined end goal.

Paradoxically, things that need either very quick resolution as well as things that are in the farther future place relatively low demand on democratic legitimation. Issues that need quick resolution must be resolved through the mobilisation of expert knowledge, and these issues are unfeasible to bring up for political debate, voting and elections, etc. Similarly undemanding but for other reasons are things in the far future: while they can certainly be availed for debate and other forms of politicisation, they are at the same time mostly not that consequential in the present, which implies that their burden of justification (through debate, representation or any other democratic mechanism) is less cumbersome.

Democratic legitimacy, and hence the relevance of contemporary notions of governance that are generally thought to complement conventional notions of democratic government, seems to be most delicate in those temporal frames that correspond to the paces and rhythms of social life. Issues that need resolutions on a time frame between months and years are most likely to attract attention, to become an issue and attract a public (Marres, 2009), and to find their way into political institutions with some perceived relevance. For these reasons, it is also at this scale that the concept of governance plays out. The latter being the recognition that much of what counts as political decisions are taken outside formalised institutions and with more complex justifications and legitimations than typically derive from political theory, makes that learning and iteration are most substantial.

Clearly, there is a difference between the time scale at which issues occur and need resolution, and the time scale at which people are actually engaged with this problem. Network instability might require action on the time scale of seconds or minutes, but the capacity to accommodate such urgent issues is built up over years or even decades. At the same time, even in this light it remains unclear how enrolling a broader political community into the management of such issues can be of any help, as the problem remains fairly expert-driven. The same holds for the far future. In far-future explorations, of course input can be sought from the widest possible range of publics, but this is not the same 'learning' as referred to here, which is rather the circulation of ideas through political, expert and public forums and practices so as to maximise the alignment of theory and practice. Thus, it is indeed the middle range where governance becomes a relevant concept of organisation.

9.4.2 Different Modes Have Different Spatial Jurisdictions

Also the spatial differentiation offered by the systemic approach to ES shows connections to modes of governance. Clearly, whether an institution of governance, whether formal or informal, has something to contribute to the resolution of a particular issue, strongly depends on whether the institution has any means of exercising power in a way that matters to the issue. Geopolitical factors are mostly beyond the reach of actors in ES issues. Similarly, issues that occur at the level of citizens and their private sphere cannot straightforwardly be put up for politicisation: the maintenance of the private sphere is one of the cornerstones of the organisation of contemporary societies.

9.4.3 Focus on Services Helps

Interestingly, the focus on services, or in a broader sense the focus on 'package deals' rather than the provision of single forms of energy, corresponds to a more integrative view of governance. For one thing, it implies an integrative definition of the policy problem. But also does it connect publics and markets, and offers a mechanism of connection between different stakeholders. The broader the range of enrolled stakeholders, the broader the range of interests that can be voiced and the broader the material available for learning and iteration are.

9.5 CONCLUSION

Most importantly, the systemic approach imposes a structure to the governance problem of ES without unduly simplifying it. With its internal richness, the systemic approach allows for more substantive recommendations on how the 'extended peer community' in heterogeneous knowledge production is to be shaped, as well as for a more substantive idea of how such knowledge can speak saliently to policy.

Governance of messy and complex processes such as energy transitions needs to impose some structure to the problem before it can address the problem in the first place. As literature on transitions in particular and on governance of complex problems in general has made clear, such structuring can only to a very limited extent be done beforehand. The vast part of it is to be 'learned' in cycles of iteration. The approach in this chapter has specified an initial direction of the structuring, after which the learning part is to unfold. It suggests that there is something special to energy transitions; that there is something that grants them more structure than just any other wicked problem. To be fair, this structure does not follow from any particular empirical reality we have observed but from the various forms of stability we have normatively defined as together covering the relevant space of stability.

In cycles of learning, reduction of both normative and empirical uncertainty is sought. Contrary to problems that are less complex and more structured by their own nature, wicked problems typically resist dissection into normative and empirical underlying uncertainties. Instead, normative and empirical uncertainties need to be addressed in tandem. This entails that multiple sorts of expertise need to be enrolled at once, and that multiple modes of governance need to be deployed.

REFERENCES

Barrett, M., Bradshaw, M., Froggatt, A., Mitchell, C., Parag, Y., Stirling, A., Watson, J., Winzer, C., 2010. Energy Security in a Multi-polar World. Discussion Draft ESMW Research Cluster. Exeter University, Exeter.

Bertel, E., 2005. Nuclear energy and the security of energy supply. NEA News 23 (2), 4–7.

Bohi, D.R., Toman, M.A., 1996. The Economics of Energy Security. Kluwer Academic Publishers, Dordrecht.

Bolado-Lavin, R., Gracceva, F., Zeniewski, P., Zastera, P., Vanhoorn, L., Mengolini, A., 2012. Best Practices and Methodological Guidelines for Conducting Gas Risk Assessments, European Commission, Joint Research Centre, Institute for Energy and Transport, JRC Scientific and Technical Report. Publications Office of the European Union, Luxembourg.

Brown, M.B., 2009. Science in Democracy: Expertise, Institutions, and Representation. The MIT Press, Cambridge MA.

Chaudry, M., Ekins, P., Ramachandran, K., Shakoor, A., Skea, J., Strbac, G., Wang, X., Whitaker, J., 2009. Working Paper. Building a Resilient UK Energy System, vol. 23. United Kingdom Economic Research Council, London.

Checchi, A., Behrens, A., Egenhofer, C., 2009. Long-term Energy Security Risks for Europe: A Sector-Specific Approach, CEPS Working Document, 309.

Cherp, A., Jewell, J., 2011. Measuring energy security: from universal indicators to contextualised frameworks. In: Sovacool, B.K. (Ed.), The Routledge Handbook of Energy Security. Routledge, Abingdon, New York.

Chester, L., 2010. Conceptualising energy security and making explicit its polysemic nature. Energy Policy 38, 887–895.

Creti, A., Fabra, N., 2007. Supply security and short-run capacity markets for electricity. Energy Economics 29 (2), 259–276.

Damigos, D., Tourkolias, C., Diakoulak, D., 2009. Households' willingness to pay or safeguarding security of natural gas supply in electricity generation. Energy Policy 37, 2008–2017.

Dancs, A., 2008. The military cost of securing energy, 2008. National Priorities Project, Inc.

Department for Business, Enterprise and Regulatory Reform (DBERR), 2007a. Expected Energy Unserved: A Quantitative Measure of Security of Supply.

Department for Business, Enterprise and Regulatory Reform, 2007b. Expected Energy Unserved. Contribution to the Energy Markets Outlook Report. Department for Business, Enterprise and Regulatory Reform, London.

Delucchi, M.A., Murphy, J.J., 2008. US military expenditures to protect the use of Persian Gulf oil for motor vehicles. Energy Policy 36, 2253–2264.

Department of Trade and Industry (DTI), 2002. Joint Energy Securityof Suppl Working Group (JESS) First Report.

Doorman, G.L., Kjølle, G., Uhlen, K., Huse, E.S., Flatabø, N., 2006. Vulnerability analysis of the Nordic power system, power systems. IEEE Transactions 21 (1), 402–410.

European Commission, 2010. Energy 2020 a Strategy for Competitive, Sustainable and Secure Energy. Brussels, 10.11.2010, COM (2010) 639.

European Commission, 2013a. Delivering the Internal Electricity Market and Making the Most of Public Intervention. Brussels, 5.11.2013, C(2013) 7243 final.

European Commission, 2000. Towards a European Strategy for the Security of Energy Supply. Green paper. COM (2000) 769 final. Brussels.

European Commission, 2006. A European Strategy for Sustainable, Competitive and Secure Energy. Green paper. COM (2006) 105 final. Brussels.

European Commission, 2013b. A 2030 Framework for Climate and Energy Policies. Green paper. COM (2013) 169 final. Brussels.

Flouri, M., Karakosta, C., Doukas, H., Flamos, A., 2008. Risks on energy security of supply: an exploratory analysis for the researcher. In: Risks on Energy Security of Supply: An Exploratory Analysis for the Researcher, Based on the Proceedings of the 1st International Conference (1st IC) of the FP7 Project: "Risk of Energy Availability: Common Corridors for Europe Supply Security" (REACCESS)", Held in Turin, Italy on the 29th of February 2008.

Gnansounou, E., 2008. Assessing the energy vulnerability: case of industrialised countries. Energy Policy 36, 3734–3744.

Gracceva, F., Zeniewski, P., 2014. A systemic approach to assessing energy security in a low-carbon EU energy system. Applied Energy 123, 335–348.

Gracceva, F., Zeniewski, P., 2015. Between Climate Change and Energy Security Objectives: The Potential Role of Shale Gas, Chapter of the Handbook for Clean Energy Systems. Wiley.

Greenleaf, J.R.H., Angelini, T., Green, D., Williams, A., Rix, O., Lefevre, N., 2009. Analysis of Impacts of Climate Change Policies on Energy Security. Ecofys, Utrecht.

Grubb, M., Butler, L., Twomey, P., 2006. Diversity and security in UK electricity generation: the influence of low-carbon objectives. Energy Policy 34 (18), 4050–4062.

Helm, D., 2003. Energy, the Market and the State: British Energy Policy Since 1979. Oxford University Press.

Hirschhausen, C., Neumann, A., Ruester, S., Auerswald, D., 2008. Advice on the Opportunity to Set Up an Action Plan for the Promotion of LNG Chain Investments – Economic, Market, and Financial Point of View – FINAL REPORT, Study for the European Commission, DG-Tren, Contracting Party. MVV Consulting, Dresden.

Huntington, H.G., 2005. EMF 23 Scenario Design, EMF WP 23.1. June 2005 (Revised October 2005).

International Energy Agency, 2001. Towards a Sustainable Energy Future. Paris.

Intharak, N., Julay, J.H., Naknishi, S., Matsumoto, T., Mat Sahid, E.J., Ormeno Aquino, A.G., Aponte, A.A., 2007. A Quest for Energy Security in the 21st Century. Asia Pacific Energy Research Centre. IEA. 2009.

de Joode, J., Kingma, D., Lijesen, M., Mulder, M., Shestalova, V., 2006. Energy Policies and Risks on Energy Markets. A Cost-Benefit Analysis. Netherlands Bureau for Economic Policy Analysis.

Jansen, J.C., 2009. Energy Services Security: Concepts and Metrics. ECN-E–09–080. Energy Research Centre of the Netherlands, Petten.

Joskow, P., 2005. Supply security in competitive electricity and natural gas markets. In: Utility Regulation in Competitive Markets. Edward Elgar Publishing Ltd.

Kaderják, P., 2011. Security of Energy Supply in Central and South-East Europe. Corvinus University of Budapest, Regional Centre for Energy Policy Research, Budapest.

Keppler, J.H., 2007a. International Relations and Security of Energy Supply: Risks to Continuity and Geopolitical Risks. University of Paris-Dauphine.

Keppler, J.H., 2007b. Energy interdependence in a multi-polar world: towards a market-based strategy for safeguarding European energy supplies. Reflets et perspectives de la vie économique XLVI (4), 31–48.

Kruyt, B., Van Vuuren, D.P., De Vries, H.J.M., Groenenberg, H., 2009. Indicators for energy security. Energy Policy 37 (6), 2166–2181.

Lefèvre, N., 2007. Energy Security and Climate Policy. Paris.

Lesbirel, S.H., 2004. Diversification and energy security risks: the Japanese case. Japanese Journal of Political Science 5 (1), 1–22.

Lieb-Doczy, E., Borner, A.R., MacKerron, G., 2003. Who secures the security of supply? European perspectives on security, competition, and liability. The Electricity Journal 16 (10), 10–19.

Marres, N., 2009. Testing powers of engagement. Journal: European Journal of Social Theory 12 (1), 117–133.

McCarthy, R.W., Ogden, J.M., Sperling, D., 2007. Assessing reliability in energy supply systems. Energy Policy 35 (4), 2151–2162.

Mitchell, J.V., 2009. Europe's Energy Security After Copenhagen: Time for a Retrofit? Chatam House Briefing Paper, Energy, Environment and Resource Governance EERG BP 2009/05.

de Nooij, M., Koopmans, C., Bijvoet, C., 2007. The value of supply security: the costs of power interruptions: economic input for damage reduction and investment in networks. Energy Economics 29 (2), 277–295.

de Nooij, M., 2011. Social cost-benefit analysis of electricity interconnector investment: A critical appraisal. Energy Policy 39, 3096–3105.

National Economic Research Associates, 2002. Security in Gas and Electricity Markets. Final Report for the Department of Trade and Industry. NERA Economic Consultants, London.

NEA/OECD, 2010. The Security of Energy Supply and the Contribution of Nuclear Energy.

O'Leary, F., Bazilian, M., Howley, M., Ó Gallachóir, B., 2007. Security of Supply in Ireland. EPSSU, Dublin.

Oxera, 2007. An Assessment of the Potential Measures to Improve Gas Security of Supply – Report Prepared for the DTI. Oxford, UK.

POYRY, 2010. Security of Gas Supply: European Scenarios, Policy Drivers and Impact on GB. A Report to Department of Energy and Climate Change. Oxford, UK.

Rutherford, J.P., Scharpf, E.W., Carrington, C.G., 2007. Linking consumer energy efficiency with security of supply. Energy Policy 35 (5), 3025–3035.

Sovacool, B., Mukherjee, I., 2011. Conceptualizing and measuring energy security: a synthesized approach. Energy 36.

Stern, J., 2004. UK gas security: time to get serious. Energy Policy 32 (17), 1967–1979.

Stern, J., 2002. Security of European Natural Gas Supplies. The Royal Institute of International Affairs.

Stern, R.J., 2010. United States cost of military force projection in the Persian Gulf, 1976–2007. Energy Policy 38 (6), 2816–2825.

Stirling, A., 1999. On Science and Precaution in the Management of Technological Risk, Volume I – A Synthesis Report of Case Studies, an ESTO Project Report Prepared for the European Commission. JRC Institute Prospective Technological Studies Seville.

Stirling, A., 2009. The dynamics of security. Stability, durability, resilience, robustness energy security in a multipolar world. In: Presentation to Workshop on 'Energy Security in a Multipolar World', 2nd April 2009. Kohn Centre, Royal Society, London.

Stirling, A., 2010. Multicriteria diversity analysis: a novel heuristic framework for appraising energy portfolios. Energy Policy 38 (4), 1622–1634.

Tosato, G., 2008. Energy security from a systems analysis point of view: introductory remarks. In: Proceedings of the 1st International Conference of the FP7 Project REACCESS. Turin.

Watson, J., 2009. Is the Move Toward Energy Security at Odds with a Low-carbon Society? In: Giddens A.,

Watson, J., 2010. UK Gas Security: Threats and Mitigation Strategies. University of Sussex.

Weare, C., 2003. The California Electricity Crisis: Causes and Policy Options. Public Policy Institute of California, San Francisco.

Winzer, C., 2012. Conceptualizing energy security. Energy Policy 46, 36–48.

Wright, P., 2005. Liberalisation and the security of gas supply in the UK. Energy Policy 33 (17), 2272–2290.

Chapter 10

Towards a Low-Carbon, Citizens-Driven Europe's Energy Security Agenda

P. Lombardi[1], B. O'Donnell[2]

[1]*Interuniversity Department of Regional and Urban Studies and Planning (DIST), Politecnico di Torino, Turin, Italy;* [2]*Ecologic Institute US, Washington, DC, United States*

10.1 REFRAMING THE DOMINANT DISCOURSE ON ENERGY SECURITY IN EUROPE

In challenging the energy security paradigm and habituated business-as-usual perspectives used to shape energy policy, this book has sought to reframe public debate and discourse surrounding Europe's energy security agenda. By positing energy security as a fundamental component of comprehensive energy policy, rather than segmenting it as a temporal (emergency) reaction to current (or past) social and political constellations, a fully integrated, resilient European energy system that adapts to climate change mitigation commitments and considers cultural and regional variations in behaviour becomes conceivable.

To get from conception to implementation, a selection of underlying issues have been analysed in the previous chapters, including the following:

1. The need to consider climate change and energy security objectives as integrated components of an energy system that is both low-carbon and secure and the need to improve frameworks for evaluating the co-benefits of climate policies.
2. Energy security should be considered through a geopolitical perspective, especially as it pertains to economics, resource competition and availability. In this regard, the significant conflict potential between the European Union (EU) and Russia has been addressed, considering the EU's current dependent position on Russian gas imports (and Russia's economic dependence on gas exports) in the short and medium terms and its stated objective of importing little or no gas from Russia in the long term (these time horizons are not always explicitly defined).
3. Large-scale renewable energy projects could play a significant role in Europe's transition to a low-carbon energy system – however, questions as

Low-carbon Energy Security from a European Perspective. http://dx.doi.org/10.1016/B978-0-12-802970-1.00010-3
Copyright © 2016 Elsevier Ltd. All rights reserved.
257

to the beneficial impacts on energy security remain. Perhaps the greatest contribution to low-carbon energy security that these mega-projects offer is the opportunity to rethink how an integrated and interdependent energy system best functions both intra- and extra-EU, along with other existing policy paradigms that continue to hinder progress in the low-carbon energy transition.

4. The complexity of energy transitions and the ambiguity of energy security as a consequence of the existence of multiple perspectives must be acknowledged. Energy security is seen as a final output of the energy transition towards a low-carbon society. This transition has to be considered as a process that is influenced by the interaction of multiple intended and unintended elements. This is partly attributable to individual attitude, in particular by the complex range of factors that influence the ways consumers operate in the current market. On the one hand, individuals are influenced by emotions, habits and the behaviour of those around them, as well as by structural factors such as transport infrastructure. Societies, on the other hand, are influenced by technological, political, economic, environmental, lifestyle and cultural factors as well as the interrelation between them and the sum of individuals that compose them. To disentangle this complexity, the book has proposed to integrate different perspectives in approaching societal energy transitions, including:

   a. The *political perspective* focused on the political processes involved in energy transition, including dynamics such as decision-making, adoption of political standards, fund raising and management, power relations between political and administrative levels.

   b. The *geopolitical perspective* which took into account the relationships between the EU and other geographic and political macroregions, exploring existing and emerging spatial and scalar alliances and dependencies.

   c. The *economic perspective* which allowed the study of the economic factors which influence societal transition, such as optimal level of prices, taxes, subsidies, resources availability and efficiency, external costs and market stability, but also the level of life satisfaction and citizen wellbeing.

   d. The *technological perspective* which characterises energy transition as a process of technology development, innovation and transfer, centred on the enhancement of sustainable energy technologies within a particular national or local context.

   e. The *environmental perspective* put forward the environmental factors that influence energy transitions, including how a society perceives the environmental impacts of energy use and a low-carbon strategy.

   f. The *lifestyle and cultural perspective* focused on lifestyle and culture influences on energy transitions (eg, social norms, societal priorities, customs, cultural needs, etc.).

Although at a European level, it has generally been recognised that citizens, communities and societal organisations are to be mobilised towards behavioural change, collected evidence shows that the ways in which such change of behaviour is to take place remains a largely disputed issue.

10.2 THE RISE OF THE HUMAN FACTOR

The research studies presented in this book largely derive from the recent EU MILESECURE-2050 project, *Multidimensional Impact of the Low-carbon European Strategy on Energy Security, and Socio-Economic Dimension up to 2050 perspective*. This showed that behaviour can only be changed if the larger context of the behaviour is understood and addressed. This requires people's empowerment through knowledge and stimulation to recognise problems as their own, all within institutional contexts ready to welcome local initiative.

To do so, the array of knowledge to be mobilised should be much broader than at present in order to match the high complexity and heterogeneity of energy transition. First, the array needs to incorporate the knowledge of citizens, which is highly complementary to formalised knowledge currently available in institutions. This knowledge is indispensable if part of the change is to take place in the households, local organisations and communities in which citizens live. At the same time, formalised knowledge should be used not only to inform policy-making but, most important, as thought-provoking material upon which debate can be conducted, behaviours shaped and human social potentials activated. The more perspectives upon a problem are mobilised, the larger the creative pool from which solutions can be tapped.

For example, it has been observed that grass-roots political and social movements are among the primary drivers of societal change. Technological, financial and social infrastructure facilitation should be in place to help scale up locally developed best practices. However, these are different areas of governance that require different governance strategies. Since infrastructures are most effectively determined by central actors, the most important principle is that they should not hamper the human energy infrastructure. Local initiatives, on the contrary, are the affairs of local communities. At this level, governance should empower local players by sharing relevant knowledge and by seeking sites in which a broader audience can be engaged (Valkenburg et al., 2015).

The book also claims that the role of citizens and communities has been, so far, inadequately understood, especially by policy-makers. Consequently, people have been insufficiently enrolled, engaged and mobilised as a resource for innovation and change in low-carbon and secure energy transitioning. In civil society, governance strategies should be aimed at acquiring insight into how citizens, communities and social groups define the problems and challenges of energy transitions.

Simultaneously, top-down approaches need to continue to play a crucial role by building social and technical infrastructures in such a way that small-scale experiments with alternative configurations for low-carbon and secure energy are enabled rather than closed off.

Both top-down and bottom-up strategies should be made explicitly complementary. Top-down interventions must be geared towards enabling local initiatives and towards empowering communities, and bottom-up initiatives might appeal to central governments so as to improve centralised regulatory regimes.

10.3 SOME MAJOR RECOMMENDATIONS FOR IMPROVING EUROPEAN ENERGY POLICIES

Building on the elements mentioned earlier as well as on the UN-COP21 Paris Agreement (United Nations, 2015), a number of guidelines and recommendations have been provided for improving European policies (Cotella et al., 2015), including each of the following:

- Energy production and provision strategies should address a diversification of sources and suppliers and a greener energy mix. Most important, policies aiming at such diversification should always take into account the constant evolution of geopolitical and geo-economic scenarios in order to always be aware of the potential risks they may bring along in terms of energy security.
- Energy efficiency improvement is a sustainable path to limit consumption and therefore dependency.
- Working towards a limited dependency from external sources, actors and investors in the energy field can contribute to provide the EU with a sufficiently strong and secure energy production system.
- Every legislation and regulation that could contribute to harmonise the EU Member State usage of energy sources will decrease or at least mitigate energy dependency.
- It is crucial that EU international diplomacy and policies firmly and proactively address the application of CO_2 reduction targets and CO_2 global prices to the wider extent of foreign countries. Political actions in this direction may have a strong impact not only on the global and European environmental situation, ie, on the mitigation of climate change, but also imply direct economic benefits, eg, in terms of profitability of EU industries on the global market.
- EU energy policy should reflect climate change policy objectives and, at the same time, take into account energy security comprehensively, ie, in terms of energy production, provision, distribution and consumption. Only in this way will it be possible to exploit the existing synergies between low-carbon transitions and energy security, as well as acknowledge and limit the negative impacts of potential contradictions between these two dimensions.
- A dimension of energy security which should be taken into account is the continued ability to deliver stable and affordable energy services to consumers

in each of the Member States, independent of the policy measures defined at the EU level.

- The EU should continue to firmly pursue the consolidation of an EU harmonised energy market. This consolidation should also provide the conditions and mechanisms to integrate local production initiatives as a prerequisite for the development of a societally driven low-carbon transition.
- Whereas an EU single energy market can provide the several benefits in the medium to long term, it needs careful planning. This is especially true in relation to its short-term impact on 'weaker' Member States' energy markets.
- In the process of creating a single energy market, the EU needs to take into account national specifications regarding energy production and available energy sources.

Complementary recommendation to be implemented at the local level include the following:

- Policymakers need to understand the role of social, political and grass-roots factors as preconditions, triggers and impact catalysers for the low-carbon energy transition. This implies significantly less emphasis on technology and on top-down planning and more emphasis on the promotion of individuals and social groups to articulate themselves and participate in the decision-making process.
- Some help will be necessary for advanced and successful local Anticipatory Experiences to be reproduced on a broader scale and become eventually mainstream. Policy-makers have significant influence on the transition process via the creation of a favourable governance and legal framework, fiscal and other incentives and regulation.
- The long-term success of the low-carbon energy transition ultimately depends on a change of personal choices and, thus, changed behaviour. This implies a transition at the personal level which cannot be forced by policy-makers, but which can be supported by soft measures in information, communication and awareness raising, but also by supporting activities focussing on the human factor.
- Whereas it is widely accepted that energy transitions cannot proceed without involving the public, this usually translates into policy measures aiming at education and at the diffusion and discussion of existing information. In order to strengthen the knowledge capital that is actually available, a more explicit public involvement is recommended, aiming at the exchange of perspectives and the joint definition of problems. This would contribute to activate and involve citizens in the process of defining problems, challenges and solutions.
- Since local knowledge is of vital importance for the achievement of sustainability goals, top-down approaches must be supportive of local initiatives in order to enhance this potential. Hence, top-down governance must be aimed at incorporating such local knowledge through a serious enrolment

of local actors into decision-making and implementation processes. This all means that top-down approaches must be critically assessed and arranged such that they become, in fact, reciprocal forms of communication instead of the classic univocal instruction.

- Even if many of the proposed policy implications largely transcend formal institutions of politics and policy-making, the latter still bear crucially important responsibilities to support such mechanisms. It is vital to keep politics inclusive and receptive, not only to stakes and interests, but also to problem-framing. While the quality of governance can be thought to be independent from particular political visions or programmes, it is likely that such inclusive views lead to more prosocial arrangements and a greater attentiveness to distributive justice, at the level of citizens as well as between Member States.

- Energy transitions are situated in current high-carbon societies, against the background of vested interests and influential societal structures. Currently, European contexts are insufficiently geared towards empowering individuals and facilitating local initiatives. Investments should not only be made directly into the energy infrastructure, but also more broadly into the context, such that citizens can play a more prominent role in dealing with energy challenges. This is about empowering citizens by establishing two-way communication channels that, on the one hand, provide them with information and resources and, on the other hand, allow for a full use of citizen knowledge.

These recommendations open the door to new collaborations, new ways of thinking and, ideally, new paradigms that will produce the policy initiatives necessary to create a low-carbon secure energy future in Europe. These new paradigms, however, will also become static, trapped in a cycle of redundancy and ineffectiveness, leading to eventual apathy from society. It will then be time to return to these recommendations, to these processes, to rejuvenate and reawaken the creative possibilities of democratic societies. Democracy is much like renewable energy: they are both self-regenerative applications of existing potentialities; they are both networks of interdependency, rather than vulnerable co-dependencies; and they are both made more secure through increased participation and constant evolution. Low-carbon energy security is the democratised evolution of Europe's energy system.

ACKNOWLEDGEMENTS

The authors wish to thank and acknowledge the contribution of all the MILESECURE-2050 project partners in achieving the results presented in this book (www.milesecure20150.eu) and Federica Borio for her support during the reviewing process.

REFERENCES

Cotella, G., Lombardi, P., Toniolo, J., 2015. MILESECURE-2050-Deliverable 5.2. Guidelines and Recommendations for European Policies. Brussels, 22-12-15. Available at: http://www.milesecure2050.eu/en/public-deliverables.

United Nations, 2015. Conference of the Parties Twenty-First Session, Framework Convention of Climate Change, FCCC/CP/2015/L.9. Paris, 12 December 2015. Available at: http://unfccc.int/resource/docs/2015/cop21/eng/l09.pdf.

Valkenburg, G., Bijker, W.E., Swierstra, T.E., 2015. MILESECURE-2050-Deliverable 5.1 Secure and Low-Carbon Energy Is Citizens' Energy. A Manifesto for Human-Based Governance of Secure and Low-Carbon Energy Transitions. Of the Project. December 2015. Maastricht University/Politecnico di Torino. Available at: http://www.milesecure2050.eu/documents/public-deliverables/en/deliverable-5-1-manifesto-for-a-governance-of-energy-transition.

Index